Smart Innovation, Systems and Technologies

Volume 348

The Smart Innovation, Systems and Technologies book series encompasses the topics of knowledge, intelligence, innovation and sustainability. The aim of the series is to make available a platform for the publication of books on all aspects of single and multi-disciplinary research on these themes in order to make the latest results available in a readily-accessible form. Volumes on interdisciplinary research combining two or more of these areas is particularly sought.

The series covers systems and paradigms that employ knowledge and intelligence in a broad sense. Its scope is systems having embedded knowledge and intelligence, which may be applied to the solution of world problems in industry, the environment and the community. It also focusses on the knowledge-transfer methodologies and innovation strategies employed to make this happen effectively. The combination of intelligent systems tools and a broad range of applications introduces a need for a synergy of disciplines from science, technology, business and the humanities. The series will include conference proceedings, edited collections, monographs, handbooks, reference books, and other relevant types of book in areas of science and technology where smart systems and technologies can offer innovative solutions.

High quality content is an essential feature for all book proposals accepted for the series. It is expected that editors of all accepted volumes will ensure that contributions are subjected to an appropriate level of reviewing process and adhere to KES quality principles.

Indexed by SCOPUS, EI Compendex, INSPEC, WTI Frankfurt eG, zbMATH, Japanese Science and Technology Agency (JST), SCImago, DBLP.

All books published in the series are submitted for consideration in Web of Science.

Srikanta Patnaik · Roumen Kountchev ·
Yonghang Tai · Roumiana Kountcheva
Editors

3D Imaging—Multidimensional Signal Processing and Deep Learning

Multidimensional Signals, Video Processing and Applications, Volume 2

Springer

Editors
Srikanta Patnaik
I.I.M.T.
Bhubaneswar, Odisha, India

Yonghang Tai
Yunnan Normal University
Kunming, China

Roumen Kountchev
Technical University of Sofia
Sofia, Bulgaria

Roumiana Kountcheva
TK Engineering
Sofia, Bulgaria

ISSN 2190-3018 ISSN 2190-3026 (electronic)
Smart Innovation, Systems and Technologies
ISBN 978-981-99-1144-8 ISBN 978-981-99-1145-5 (eBook)
https://doi.org/10.1007/978-981-99-1145-5

This Springer imprint is published by the registered company Springer Nature Singapore Pte Ltd.
The registered company address is: 152 Beach Road, #21-01/04 Gateway East, Singapore 189721, Singapore

Preface

This is Volume 2 of the Proceedings of the Fourth Conference on 3D Imaging Technologies—Multidimensional Signal Processing and Deep Learning (3D IT-MSP&DL).

The 3D Imaging Technologies attracted recently significant attention both in research and in industry, and the topics cover many related aspects of Multidimensional Signal Processing, Deep Learning and Big data. The 3D IT-MSP&DL'22 provided a wide forum for researchers and academia as well as practitioners from industry to meet and exchange their ideas and recent research works on all aspects of multidimensional signals analysis and processing, applications and other related areas.

The large number of conference topics attracted researchers working in various scientific and application areas. The papers, accepted for publishing in the proceedings, are arranged in two volumes.

The selection of papers in Volume 2 covers research works presenting new theoretical approaches and related applications in wide variety of areas, such as: sentiment analysis; time series forecasting; development of test methods and equipment for intelligent cockpit; automated inspection system; data analysis model; self-supervised representation by pretext task; quantitative control technology; fungus detection; registration between SAR and optical images; strategy for battery energy storage; digital power grid architecture; vessel image segmentation; recognition of license plates; food safety system; fault diagnosis of chemical processes; satellite spatial temperature measurement; plant allocation of sponge city; interactive power distribution technology; color recognition of plants; underwater fishing technology; network traffic classification; flipped classroom teaching; furniture design; and some specific mathematical tasks with related applications in multidimensional data analysis. In their investigations, authors used various approaches based on deep learning, CNN, RNN, U-Net, dynamic analysis, nonlinear Gardner equation, 3D virtual reality, attention-enhanced encoder-decoder network and many other contemporary scientific tools.

The aim of the book is to present the latest achievements of the authors to a wide range of readers: IT specialists, researchers, physicians, Ph.D. students and other specialists in the area.

The book editors express their special thanks to Prof. Lakhmi Jain (Honorary Chair); Prof. Dr. Srikanta Patnaik, Prof. Dr. Junsheng Shi and Prof. Dr. D.Sc. Roumen Kountchev (General Chairs); Prof. Yingkai Liu (Organizing Chair); Prof. Dr. Yonghang Tai, Dr. Shoulin Yin and Prof. Dr. Hang Li (Program Chairs); and Dr. S. R. Roumiana Kountcheva (International Advisory Chair).

The editors express their warmest thanks to the excellent Springer team which made this book possible.

Bhubaneswar, India
Sofia, Bulgaria
Kunming, China
Sofia, Bulgaria
January 2023

Srikanta Patnaik
Roumen Kountchev
Yonghang Tai
Roumiana Kountcheva

Contents

About the Editors

Prof. Dr. Srikanta Patnaik is Director of I.I.M.T., Bhubaneswar, India.

Prof. Srikanta Patnaik has supervised more than 30 Ph.D. Theses and 100 Master theses in the area of Computational Intelligence, Machine Learning, Soft Computing Applications and Re-Engineering. Dr. Patnaik has published more than 100 research papers in international journals and conference proceedings. He is author of 3 text books and edited more than 100 books and few invited book chapters, published by leading international publisher like IEEE, Elsevier, Springer-Verlag, Kluwer Academic, IOS Press and SPIE.

Prof. Patnaik is awarded with MHRD Fellowship by the Government of India, for the year 1996. He is nominated for MARQUIS Who's Who for the year 2004 and nominated as International Educator of the Year 2005 by International Biographical Centre, Great Britain. He ahs been awarded with the certificate of merit from The Institute of Engineers (India) for the year 2004–05. He is also Fellow of IETE, Life Member of ISTE, and CSI. Dr. Patnaik has visited various countries such as Japan, China, Hong Kong, Singapore, Indonesia, Iran, Malaysia, Philippines, South Korea, United Arab Emirates, Morocco, Algeria, Thailand and Vietnam for delivering Key note addresses at various conferences and symposiums.

Prof. Dr. Roumen Kountchev D.Sc. Technical University of Sofia, Bulgaria.

Roumen Kountchev is with the Faculty of Telecommunications, Department of Radio Communications and Video Technologies—Technical University of Sofia, Bulgaria. He has 434 papers published in magazines and conference proceedings, and 22 patents. A member of Euro Mediterranean Academy of Arts and Sciences; President of Bulgarian Association for Pattern Recognition; Editorial board member of IJBST Journal Group; Editorial board member of: Intern. J. of Reasoning-based Intelligent Systems; Intern. J. Broad Research in Artificial Intelligence and Neuroscience; Editor of books for Springer SIST series, Guest editor of several Special Issues for "Symmetry" MDPI.

Prof. Dr. Yonghang Tai Yunnan Normal University, Kunming, China.

Dr. Yonghang Tai is a professor in the School of Physics and Electronic Information, Yunnan Normal University, Kunming, China, Color & Image Vision Lab. He got his M.Sc. at Yunnan Normal University, Kunming, and Ph.D. in Opto-Electronic Engineering (OE) at Deakin University, Melbourne, Australia. His main research interests are in 3D HMD design, AM-OLED drive circuit design, and Stereoscopic imaging systems. Professor Tai had published more than 60 journal and conference papers and has 6 patents. He chairs many high-level research projects, such as the NSFC and the provincial natural science foundations. Professor Tai was the editor and reviewer of many indexed journals. He was a tutor of M.Sc. and Ph.D. students who successfully defended their works.

Dr. Roumiana Kountcheva S. R. TK Engineering, Sofia, Bulgaria.

Dr. Roumiana Kountcheva is the Vice President of TK Engineering. She got her M.Sc. and Ph.D. at the Technical University of Sofia, Bulgaria, and became Senior Researcher (S. R.) in 1993. She has more than 200 publications (including 28 book chapters and 5 patents), and presented 23 plenary speeches at international conferences and workshops. She is a member of IRIEM, IDSAI, IJBST Journal Group, and Bulgarian Association for Pattern Recognition. Dr. Kountcheva is a reviewer of WSEAS conferences and journals, and edited several books for Springer SIST series, Guest editor ofseveral Special Issues for "Symmetry" MDPI.

Chapter 1
Prediction Based on Sentiment Analysis and Deep Learning

Haiyang Liu

Abstract This paper uses programs written with Python to predict stocks by combining benchmark sentiment analysis and deep learning. When the stock market is developing steadily, technical prediction is the main prediction method. Technical prediction is based on the stock data crawled by Quandl to predict the model, compare the prediction results of multiple models, and select the model with the highest fit with the stock price trend. When the stock market develops steadily, the fitting effect of technical prediction is very good, but when the stock market has a major emergency, the technical prediction cannot fit the stock price trend, and then, the benchmark prediction is the best way to predict. Benchmark prediction can predict the trend of stocks according to the information released by shareholders on social platforms in case of emergencies. The benchmark prediction method is to crawl the stock comments from [1] and screen the stock comments for false news. After the false news is screened, the comments judged to be true news are preprocessed. After the real news comments are processed into text fragments, the text fragments are subjected to sentiment analysis, and the rise and fall indicators are constructed by using the results of sentiment analysis. Finally, the rise and fall indicators are combined with stock price data to draw a graph, and the future stock price trend is predicted according to the trend of the curve in the graph. This paper improves the defect that the technical side cannot accurately predict when encountering special circumstances and greatly improves the accuracy of stock prediction in practical application.

H. Liu (✉)
Xi'an University of Technology (Qujiang Branch), No. 58, Yanxiang Road, Qujiang New District, Xi'an 710054, Shaanxi, People's Republic of China
e-mail: xautocean@163.com

S. Patnaik et al. (eds.), *3D Imaging—Multidimensional Signal Processing and Deep Learning*, Smart Innovation, Systems and Technologies 348,
https://doi.org/10.1007/978-981-99-1145-5_1

1.1 First Section

Stock price prediction has always been the most elusive thing, and people have been unable to predict how it will change. Even though people have created many technical prediction tools, such as moving average algorithm (see Chart 1.1), KNN algorithm [2] (see Chart 1.2), linear regression algorithm (see Chart 1.3), prophet algorithm (see Chart 1.4), ARIMA algorithm (see Chart 1.5), and LSTM algorithm (see Chart 1.6). The vertical coordinates are stock prices. The row coordinates are the time axis. The blue color is the historical stock price trend. Orange is the future stock price trend. Green is the program's predicted stock price trend. They still cannot successfully predict the trend of stock prices in reality.

Why is there still a deviation, even a huge deviation, in predicting stock prices? The answer is that there are many uncertainties affecting the price changes of stocks, which makes it extremely difficult to accurately predict this matter. Among these uncertain factors, the most important factor affecting stock prediction is the daily events, especially those related to economic activities, including the development of world trade, the release of policies by various countries, and the military activities of other countries. These events are sudden, unpredictable, and random, and they often recur. Therefore, if we want to predict the trend of stocks more accurately, we need to get relevant information in time and quickly and then to use the relevant information to screen, extract, and calculate the price trend and trend of the next stock.

According to the above characteristics, we need to divide the stock prediction into two parts, namely technical prediction and benchmark prediction. Technical prediction means that the whole trend of the stock continues to develop on the original development track without the occurrence of major emergencies. In this case, the historical price of the stock can be simulated by using various existing forecasting tools. Find a model with the best fit, check, and verify whether the future

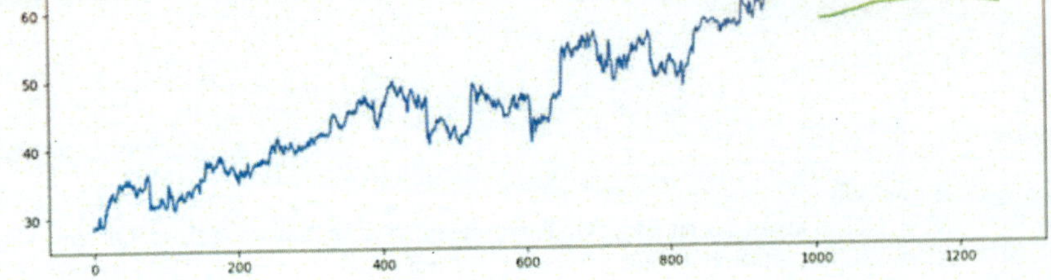

Chart 1.1 Moving average algorithm chart

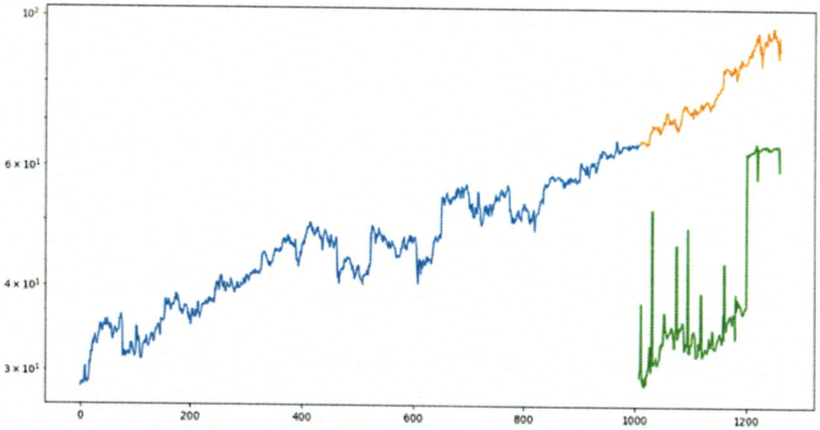

Chart 1.2 KNN algorithm chart

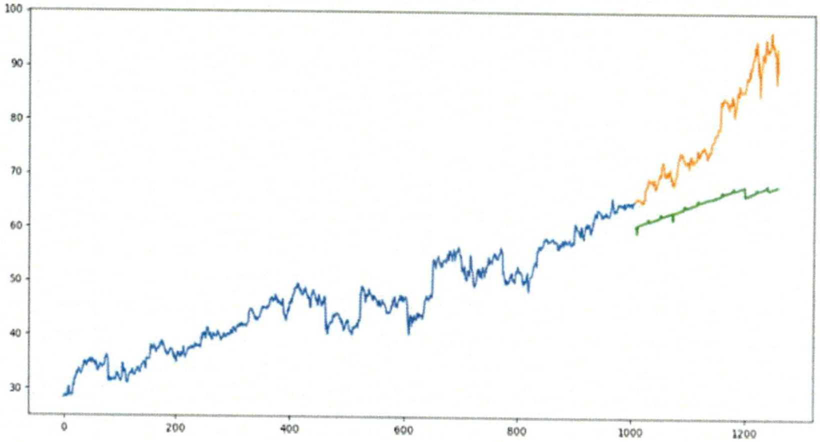

Chart 1.3 Linear regression algorithm chart

trend and trend of the stock price in the model are consistent with reality, and if they are consistent, you can judge the future trend of the stock price [3]. Benchmark prediction is based on the premise of unpredictable major events, and the stock trend has been unable to follow the original development trajectory. In this case, start the benchmark prediction and extract the information affecting the stock trend from many external factors to calculate and make the corresponding prediction.

This paper introduces benchmark prediction and how to realize the application of benchmark prediction in the real stock prediction is described in detail. This kind of prediction is very accurate in major emergencies. For example, in 2021, the stock of New Oriental evaporated 90% of its shares because of the double reduction policy issued by the state. At that time, the shareholders' mood was very low, the

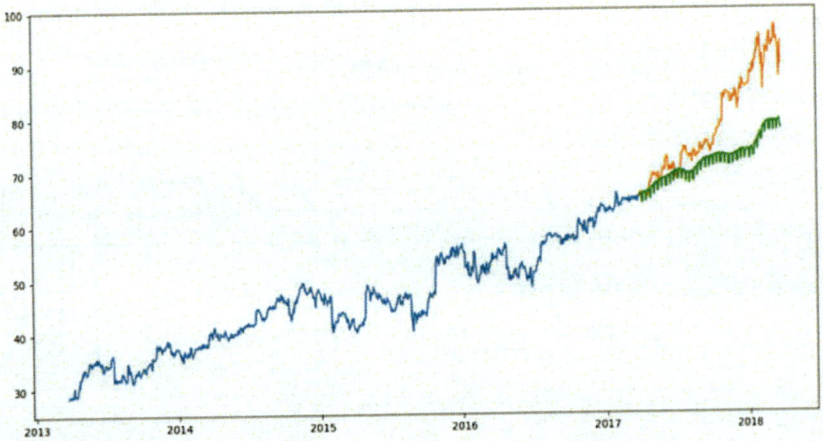

Chart 1.4 Prophet algorithm chart

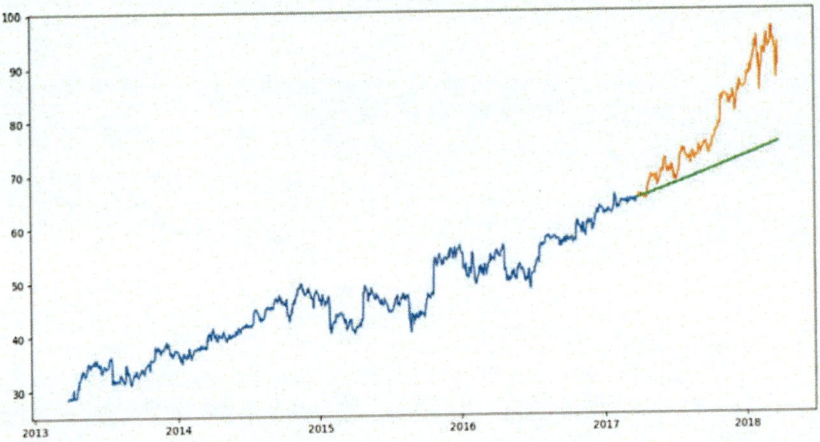

Chart 1.5 ARIMA algorithm chart

shareholders' mood index was very low, and the stock price continued to fall in the pessimistic mood of the shareholders. If we could predict the benchmark at that time, we could get the future trend of the stock price from the shareholders' mood or sell shares early to reduce the subsequent loss caused by the sharp decline in stock prices.

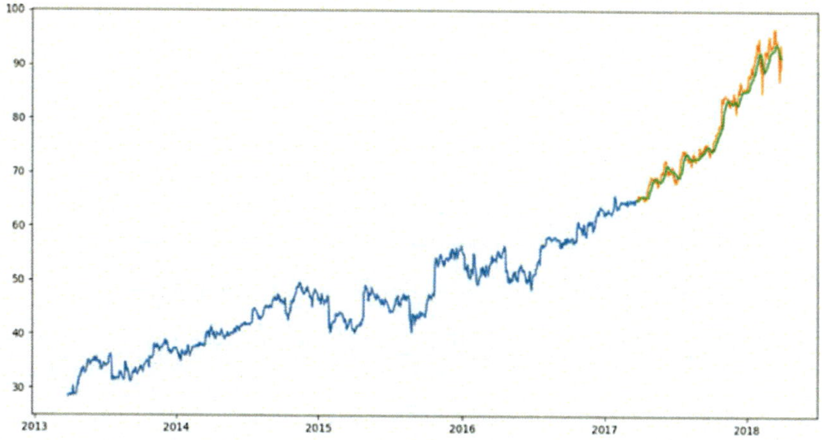

Chart 1.6 LSTM algorithm chart

1.2 Benchmark Prediction

Next, we will focus on the benchmark prediction part. Benchmark prediction is to extract the information affecting the stock trend from many external factors for information cleaning, text processing, sentiment analysis, index calculation, and trend mapping to predict the future stock price. At the end of the prediction, the stock can be traded according to the future stock price trend chart.

1.2.1 Capture Data of Stock Comments from www.guba.eastmoney.com [1]

Shareholders are a large group of Internet users, and Guba are forums for them to discuss and express their views on stocks, their mood partly reflects what is going on in the stock market, as well as the volatility of the stock market and that is why capturing stock comments is the first step to take. Entering a comment post requires two layers to find the location of the crawler, so there are two main functions in the code: get_url to get the address of the second web page, and get_comments to get the comment content.

Run get_url first. By viewing code results, we can get the structure of the comment sentence through html analysis, extract the single subject we want to parse, and then extract and append in a loop. Run get_comments., write csv, and set the writing parameters when writing. And csv can avoid error codes when writing. For example, see Table 1.1.

Table 1.1 Collected stock comments

Created_time	Title
2017/3/30 9:29	A-share speechless
2017/3/30 9:29	To and fro
2017/3/30 9:29	Now build a warehouse and wait for meat after the festival
2017/3/30 9:29	Why did A-shares plunge today?
…	…
2018/5/4 15:01	Institutional fund garbage
2018/5/4 15:01	The road ahead is confused
2018/5/4 15:01	I hope we can get off to a good start in May
2018/5/4 15:01	Raise your hand for 9 consecutive days

1.2.2 Build a Model for False News Judgment

Because the Internet is full of malicious journalists who publish news contrary to facts and logic, such as news articles that are intentional, proven wrong, and misleading to readers. They are fabricated to deceive others in order to achieve corresponding economic purposes. The information they publish is false, objective, and special purpose and usually contains strong subjectivity and incitement [4]. This will have a huge impact on the machine learning model, which will cause unpredictable and huge errors in the subsequent prediction. Therefore, before using the stock comment data captured from www.eastmoney.com [5], we should conduct a false news test to screen out the real stock comments. The false news discrimination program can extract their evaluation characteristics from different passages of news articles [6], such as the narration of real events, the description of false events, the negative description of real events, and the negative description of false events. By using Python to loop in the text to be tested from CSV, feature selection is done in FeatureSelection. For example, see Table 1.2.

Next, the prediction model is created, and the classifiers are established by Naive Bayes model [7], logistic regression [8], linear SVM [9], SVM hinge loss stochastic

Table 1.2 Training material library

Statement	Label
Says the Annies list political group supports third-trimester abortions on demand	FALSE
When did the decline of coal start? It started when natural gas took off that started to begin in (President George W.) Bushs administration	TRUE
…	…
The economic turnaround started at the end of my term	TRUE
The Chicago bears have had more starting quarterbacks in the last 10 years than the total number of tenured (UW) faculty fired during the last two decades	TRUE

gradient descent algorithm, and random forest, respectively, and then, the two models with the best performance are selected from all the installed models [10]. We call this the candidate model, and from the confusion matrix, we can see that random forest and logistic regression perform the best in terms of precision and recall. Running the random forest and logistic regression models again, we find that the n-gram forest model had better accuracy than the forest model with parameter estimation by running the random forest and logistic regression with the best parameter estimation from GridSearch. The performance of the logistic regression model with the best parameters is almost similar to that of the n-gram model, so logistic regression is the preferred prediction model. The pre-labeled false news dataset is used to train the logistic regression model. The different feature representations are weighted by aggregation method, and the weights of features are optimized.

1.2.3 Test Stock Comments for False News

The stock comment data is detected and screened by the constructed false news detection model. The stock comment text paragraphs selected from the FeatureSelection program features are judged and scored [11] in a loop with the trained prediction model, and stock comments with a score of more than 0.5 are regarded as real news.

The stock comments judged to be real news are segmented to eliminate the auxiliary words that have no semantic meaning [12]. Finally, the sentiment segments are saved to csv. For example, see Table 1.3.

Table 1.3 Comment fragments after processing

Created_time	Title
2017/3/30 9:29	Speechless
2017/3/30 9:29	To fro
2017/3/30 9:29	Build warehouse wait for meat festival
2017/3/30 9:29	Why plunge
...	...
2018/5/4 15:01	Garbage
2018/5/4 15:01	Road confused
2018/5/4 15:01	Good start May
2018/5/4 15:01	Raise hand consecutive

1.2.4 Build a Sentiment Classification Model for Stock Comment

Python machine learning model [13] is used to analyze the emotional tendency of stock comments by using the annotated corpus (see Figs. 1.1 and 1.2), and the positive and negative emotions in stock comments are classified and scored [14]. And save the classification and scoring results to csv. For example, see Table 1.4.

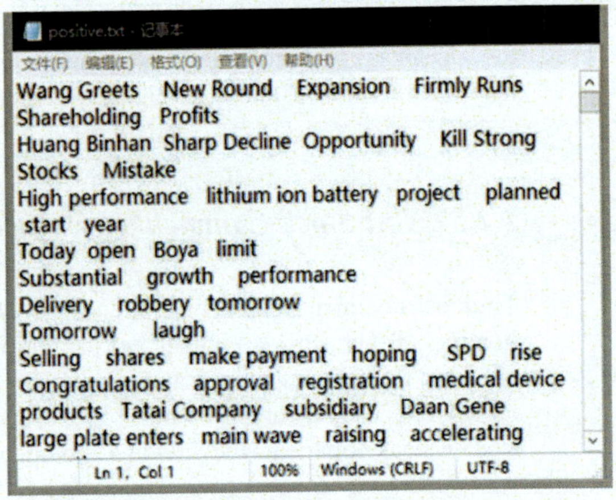
positive.txt - 记事本
文件(F) 编辑(E) 格式(O) 查看(V) 帮助(H)
Wang Greets New Round Expansion Firmly Runs
Shareholding Profits
Huang Binhan Sharp Decline Opportunity Kill Strong
Stocks Mistake
High performance lithium ion battery project planned
start year
Today open Boya limit
Substantial growth performance
Delivery robbery tomorrow
Tomorrow laugh
Selling shares make payment hoping SPD rise
Congratulations approval registration medical device
products Tatai Company subsidiary Daan Gene
large plate enters main wave raising accelerating
Ln 1, Col 1 100% Windows (CRLF) UTF-8

Fig. 1.1 Positive corpus chart

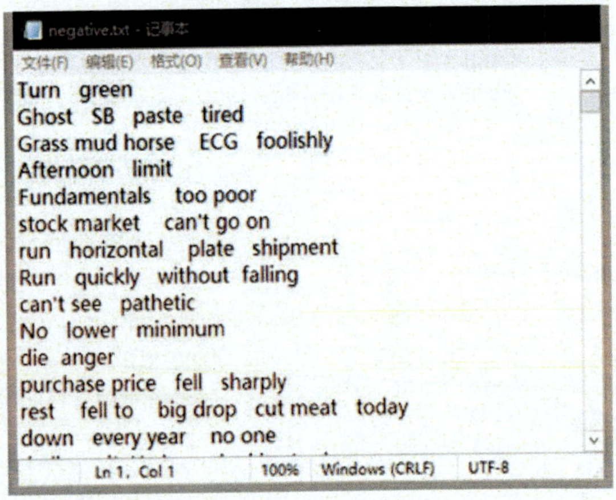
negative.txt - 记事本
文件(F) 编辑(E) 格式(O) 查看(V) 帮助(H)
Turn green
Ghost SB paste tired
Grass mud horse ECG foolishly
Afternoon limit
Fundamentals too poor
stock market can't go on
run horizontal plate shipment
Run quickly without falling
can't see pathetic
No lower minimum
die anger
purchase price fell sharply
rest fell to big drop cut meat today
down every year no one
Ln 1, Col 1 100% Windows (CRLF) UTF-8

Fig. 1.2 Negative corpus chart

Table 1.4 Classification and scoring results

Created_time	Title	Polarity
2017/3/30 9:29	Speechless	0
2017/3/30 9:29	To fro	0
2017/3/30 9:29	Build warehouse wait for meat festival	1
2017/3/30 9:29	Why plunge	0
…	…	…
2018/5/4 15:01	Garbage	0
2018/5/4 15:01	Road confused	0
2018/5/4 15:01	Good start May	1
2018/5/4 15:01	Raise hand consecutive	0

1.2.5 Build an Index from the Analysis Results

Combine the classification results and scoring results with the emotional bullish and bearish index formula to construct the rise and fall indicators.

$$BI^* = \ln\left[\frac{1 + M^{Bull}}{1 + M^{Bear}}\right] \tag{1.1}$$

1.2.6 Capture and Load Data

Python is used to get historical data for various stocks from Quandl's dataset. In this paper, I use data from "Microsoft".

Use Python to download the dataset, load the dataset, and view the first few rows of data. There are multiple variables in the dataset, such as Date = date, Open = the opening price of the day, High = the highest price of the day, Low = the lowest price of the day, Close = Closing price of the day, Volume = trading volume of the day, Ex-Dividend = excluded dividend Split Ratio = split ratio, Adj.Open = adjusted opening price, Adj.High = adjusted highest price, Adj.Low = adjusted lowest price, Adj.Close = adjusted closing price, and Adj.Volume = adjusted trading volume. (The stock market is closed on weekends and public holidays.) For example, see Table 1.5. Then, draw it into a picture (see Fig. 1.3).

1.2.7 A Subsection Sample

Image processing is carried out by using the constructed rise and fall indicators and the daily closing price data of stocks captured from Quandl. After processing, the

Table 1.5 History stock data

Date	Open	High	…	Adj.Low	Adj.Close	Adj.Volume
2013-03-27	28.14	28.445	…	24.799153	25.055270	36,047,400.0
2013-03-28	28.32	28.660	…	24.958122	25.262813	55,753,800.0
2013-04-01	28.64	28.660	…	25.046438	25.267229	29,201,100.0
2013-04-02	28.59	28.850	…	25.187744	25.435029	28,456,500.0
2013-04-03	28.75	28.950	…	25.205407	25.223070	35,062,800.0
…	…	…	…	…	…	…
2018-05-02	3087.409	3097.604	…	3064.763	3081.177	13,418,456,600
2018-05-03	3074.517	3105.66	…	3056.157	3100.859	13,998,562,000
2018-05-04	3093.117	3104.093	…	3086.785	3091.033	11,871,126,500
2018-05-07	3094.899	3136.836	…	3091.658	3136.645	13,894,818,600
2018-05-02	3087.409	3097.604	…	3064.763	3081.177	13,418,456,600

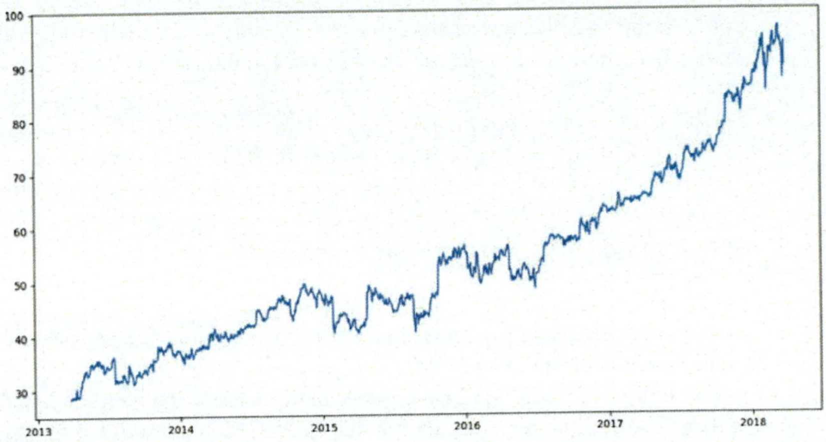

Fig. 1.3 Historical price chart

relationship between bullish and bearish sentiment and stock price movements can be drawn. We find that bullish and bearish sentiment and stock price movements are highly coincident when the stock market has a large time, and can be predicted with the same trend when the stock market develops smoothly (see Fig. 1.4).

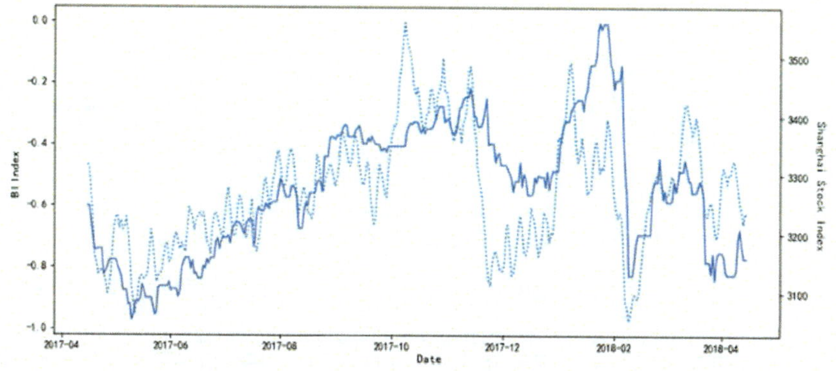

Fig. 1.4 Base level forecast price and historical price chart

1.3 Conclusion

Benchmark prediction and the traditional technical prediction can be combined to predict the real-time stock market [15]. Because benchmark prediction can analyze the sentiment index of shareholders and roughly predict the next stock price trend, it can improve the loopholes of stock application software in predicting huge deviations when unexpected events related to economic activities occur and realize the real automation of software, rather than the huge loss of users caused by the failure to adjust the prediction results when sudden, unpredictable, and random events such as politics and military occur. This technology can greatly improve the accuracy of stock prediction and make the prediction results closer to reality.

References

1. www.guba.eastmoney.com. Accessed 20 Aug 2022
2. Zhang, P.Q.: Stock price forecast based on SVM-KNN. Stat. Appl. **08**(06), 859–871 (2019)
3. Feng, C.Y., Gao, M.S.: An improved ARIMA method based on functional principal component analysis and bidirectional bootstrap and its application to stock price forecasting. Acad. J. Comput. & Inf. Sci. **5.0**(10.0) (2022)
4. Kumar, A.S., Kalpana, P., Athithya, K.S., Sundar, V.S.A.: Fake news detection on social media using machine learning. J. Phys.: Conf. Ser. **1916**(1), 012235 (2021). IOP Publishing
5. www.eastmoney.com. Accessed 20 Aug 2022
6. Ozbay, F.A., Alatas, B.: Fake news detection within online social media using supervised artificial intelligence algorithms. Phys. A: Stat. Mech. Appl. **540**(C), 123174 (2020)
7. Mustofa, R.L., Prasetiyo, B.: Sentiment analysis using lexicon-based method with naive bayes classifier algorithm on #newnormal hashtag in twitter. J. Phys.: Conf. Ser. **1918**(4), 042155 (2021). IOP Publishing
8. Nennuri, R., Yadav, M.G., Vahini, Y.S., Prabhas, G.S., Rajashree, V.: Twitter sentimental analysis based on ordinal regression. J. Phys.: Conf. Ser. **1979**(1), 012069 (2021). IOP Publishing

9. Hidayat, T.H.J., Ruldeviyani, Y., Aditama, A.R., Madya, G.R., Nugraha, A.W., Adisaputra, M.W.: Sentiment analysis of twitter data related to Rinca Island development using Doc2Vec and SVM and logistic regression as classifier. Procedia Comput. Sci. **197**, 660–667 (2022)
10. Prajval, S., Deshakulkarni, S.V.: Comparative study of various approaches, applications and classifiers for sentiment analysis. Glob. Transit. Proc. **2**(2), 205–211 (2021)
11. Imran, A.S., Daudpota, S.M., Kastrati, Z., Batra, R.: Cross-cultural polarity and emotion detection using sentiment analysis and deep learning on COVID-19 related tweets. IEEE ACCESS **8**, 181074–181090 (2020)
12. Grover, V.: Exploiting emojis in sentiment analysis: a survey. J. Inst. Eng. (India): Ser. B 1–14 (2021)
13. Goswami, A., Krishna, M.M., Vankara, J., Gangadharan, S.M.P., Yadav, C.S., Kumar, M., Khan, M.M.: Sentiment analysis of statements on social media and electronic media using machine and deep learning classifiers. Comput. Intell. Neurosci. 9194031–9194031 (2022)
14. Chen, L., Huang, Y.: Sentiment analysis of microblog comments based on multi-feature fusion. In: Conference proceedings of 2021 4th International Conference on Algorithms, Computing and Artificial Intelligence (ACAi 2021), pp. 582–586 (2021)
15. Ekaterina, L., et al.: Forecasting directional bitcoin price returns using aspect-based sentiment analysis on online text data. Mach. Learn. 1–24 (2021)

Chapter 2
A Survey on Time Series Forecasting

Xiaoxu He

Abstract Time series data are widely available in finance, transportation, tourism, and other vital fields and often reflect the dynamic pattern of the observed objects. Scientific and accurate time series forecasting can reduce system operating costs and lower system risk. However, in the era of big data, new forms of data and complex relationships among variables in the data bring significant challenges to traditional forecasting methods. In contrast, artificial intelligence methods can fully mine massive data and thus are widely used in time series forecasting problems. In this paper, machine and deep learning methods are compared and jointly applied to univariate time series prediction scenarios. Experimental results show that deep learning methods outperform machine learning methods in prediction accuracy, but their complex network structures require more training time.

2.1 Introduction

Time series is a set of data arranged in the order of the time of generation, usually the results obtained from the observation of a specific process or phenomenon at a particular time according to a specific frequency, which can reflect the process and state of the observed object over time, and this kind of data widely exists in major essential fields. Time series forecasting refers to using historical acquired time series data and the a priori knowledge gained in the area to mine and analyze the various time series patterns and then predict the outcome of future moments. By mining and analyzing time series forecasts, it is possible to optimize the allocation of resources and predict future trends based on the forecast results, as it has become a fundamental and crucial analytical tool in many fields. Currently, the primary time series forecasting methods are classified as statistical regression, machine learning (including deep learning), and combinatorial model.

X. He (✉)
Ohio State University, Columbus, OH, USA
e-mail: he.1747@osu.edu

S. Patnaik et al. (eds.), *3D Imaging—Multidimensional Signal Processing and Deep Learning*, Smart Innovation, Systems and Technologies 348,
https://doi.org/10.1007/978-981-99-1145-5_2

Typically used statistical regression methods are ARMA (Autoregressive moving average), ARIMA (Autoregressive integrated moving average), etc. Duran and Catak [1] proposed an iterative autoregressive (AR) time series model to predict wind speed. A window-shifted scheme was used to improve the performance of the algorithm. The measured and model output data were compared based on the power density spectrum. According to the calculated results, the model predicts wind speed within a 95% confidence interval. Yasmeen and Sharif [2] used the ARIMA model to predict Pakistan's electricity consumption based on data from 1990 to 2011. Using statistical regression for forecasting often gives satisfactory results in linear or low-dimensional problems. Eljazzar and Hemayed [3] used an ARIMA model and exponential smoothing to predict the grid system's overall resource utilization and response time to improve utilization and meet business demand by adjusting business system resources. However, time series data in practical applications usually exhibit nonlinear characteristics due to the combined effects of internal and external factors. Traditional models cannot learn the nonlinear features in time series, and the prediction accuracy in real applications appears not to be ideal. Therefore, artificial intelligence forecasting techniques are used to solve time series forecasting problems, which can discover the features implied in time series data and have a more powerful learning capability.

Machine learning-based forecasting models can perform well in nonlinear or high-dimensional problems by discovering and summarizing patterns and features from monitored data. Thus, typical machine learning models, including SVM (Support Vector Machine), RF (Random Forest), and DT (Decision Tree), are widely studied in the area. Lahouar and Slama [4] used a random forest regressor to predict wind energy, and the experimental results showed that this model can significantly improve the prediction accuracy compared with the traditional neural network. SVM uses the structural risk minimization principle and the Vapnik–Chervonenkis dimension theory of statistical theory to continuously optimize learning to find the optimal partition plane, which has an excellent performance in dealing with high-dimensional, nonlinear problems. Hu et al. [5] constructed an improved SVM model for short-term wind speed prediction. Lin et al. [6] applied the improved SVM model to predict financial risk and achieved good results.

In recent years, the field of deep learning developed rapidly and has become a new branch of machine learning. Many scholars have applied deep learning methods to time series forecasting and achieved satisfactory performance. One of the advantages of deep learning is that it can automatically construct more complex features than other machine learning methods at each calculation step. Moreover, deep learning has been proven to be effective in achieving high accuracy in time series prediction. Huang and Wei [7] used CNN (Convolutional Neural Networks) to solve the photovoltaic prediction problem. Zeroual et al. [8] used LSTM (Long Short-term Memory) to predict the number of newly infected by COVID-19 and recoveries. Zhang and Ci [9] used DBN (Deep Belief Networks) to forecast the gold price, and its prediction results were significantly better than those based on ARIMA and BPNN (Backpropagation Neural Networks).

The combinatorial model approach improves prediction accuracy by combining several different models to compensate for the deficiencies of a single model in specific aspects. Shi et al. [10] first predicted the original data using ARIMA, then used SVM to indicate the error derived from the previous prediction step, and combined the two predictions to obtain the final forecast value. A hybrid wind speed prediction method combining EEMD and SVM was proposed by Hu et al. [11]. The original wind speed data is decomposed into several independent and residual series using the decomposition principle. Then, the corresponding estimates are generated using the SVM algorithm. Finally, the integration principle combines these corresponding estimates into the final wind speed prediction. A case study of wind speed at three wind farms in northwestern China shows that the hybrid method is effective for unstable and irregular wind speed prediction. Darwish et al. [12] summarized the literature and found that combining techniques such as genetic algorithms and particle swarm optimization algorithms with neural networks shows excellent advantages in solving nonlinear time series forecasting problems.

This paper briefly introduces time series forecasting based on traditional machine learning and deep learning methods. We compare the two ways by applying them to real univariate time series forecasting cases.

2.2 Traditional Machine Learning-Based Method

Time series forecasting is closely associated with regression tasks in machine learning, and the execution has vast similarities. Figure 2.1 depicts the procedure of time series prediction based on traditional machine learning methods. First, a set of features related to the prediction results is extracted by feature engineering, and the corresponding training dataset and test dataset are constructed. The purpose of feature selection is to select the most beneficial features for prediction from the constructed feature set and reduce the unnecessary waste of computational resources during model training. The loss function and training data are jointly used in the model training phase to obtain the final prediction model. Once the training is completed, the model can be used for time series forecasting. In the prediction phase, the features of the existing historical data are input into the prediction model, and the prediction model gives the predicted future data.

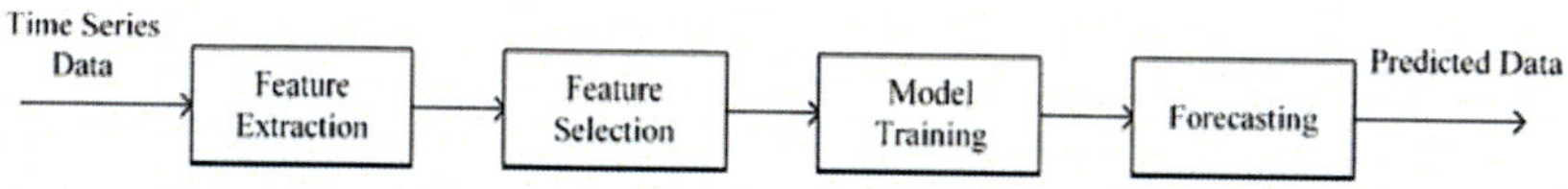

Fig. 2.1 Procedure of time series prediction using machine learning methods

Table 2.1 Partial features supported by tsfresh

Name	Illustration
abs_energy	The absolute energy of the time series which is the sum over the squared values
absolute_maximum	The highest absolute value of the time series
fft_coefficient	The Fourier coefficients of the one-dimensional discrete Fourier Transform
sum_of_reoccurring_values	The sum of all values, that are present in the time series more than once
lempel_ziv_complexity	The complexity estimate based on the Lempel–Ziv compression algorithm
matrix_profile	The 1-D matrix profile and returns Tukey's five number set plus the mean of that matrix profile

2.2.1 Feature Extraction

In machine learning, feature extraction refers to the process of creating new features from an initial set of data, which can be used to provide a more representative subset of input variables. These features encapsulate the central properties of a data set and represent it in a low-dimensional space that facilitates the learning process.

Tsfresh is one of the python packages, and it performs a huge number of time series characteristic calculations. In addition, Tsfresh is a typical python feature extraction library for time series data, which is capable of extracting more than 70 features on a variable, covering statistical features, fitted features, time domain features, and frequency domain features. Some of the features and their definitions are shown in Table 2.1.

2.2.2 Feature Selection

The feature extraction stage collects a large number of features that are relevant or potentially relevant to the prediction object. Still, at the same time, there may be features that are not relevant or have low relevance to the prediction object. If these features are introduced into the model training phase, it will not only bring no improvement in forecasting accuracy but also increase the model training time. Meanwhile, numerous features may be redundant, i.e., different features may contain the same or highly relevant information. Feature selection aims to eliminate irrelevant or redundant components and obtain a valid subset of features. Feature selection can be made directly using the tsfresh library.

2.2.3 *Model Training*

Model training refers to the use of labeled feature data to update the parameters in the model through multiple epochs so that the forecasting model learns the mapping relationship from the feature data to the target prediction value.

2.2.4 *Rolling Time Series Forecasting*

Rolling is a technique of converting a single time series into multiple time series. Suppose one has price data on a company's stock price for the last 100 days. Removing the current day's price as the target value and extracting features for the previous 99 days constitute only one sample for model training or inference. However, the sample volume will increase if the stock price of every day in the time series is removed and features are extracted until this value to train to predict the value of the day deleted. Figure 2.2 illustrates the process of rolling time series to obtain multiple sub-series from original series.

2.3 Deep Learning-Based Method

The accuracy of forecasting by machine learning methods relies heavily on manual feature engineering. Meanwhile, deep neural networks can automate the extraction of features without manual feature selection and have achieved great success in the fields of image and speech. Numerous scholars have tried applying deep learning methods

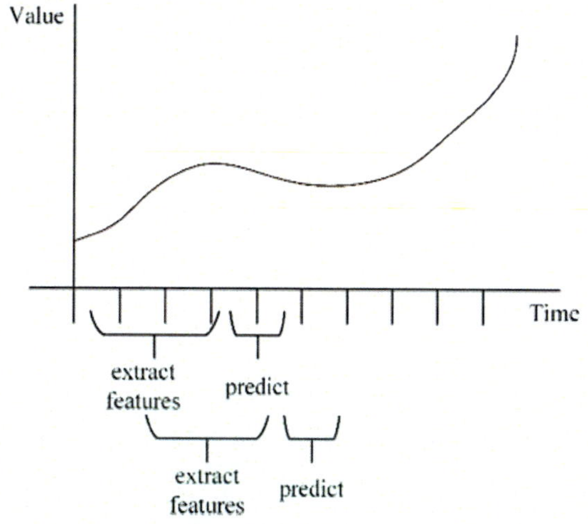

Fig. 2.2 Explanation for rolling

to time series prediction. The commonly used deep learning methods include RNN (Recurrent Neural Network), LSTM, and GRU (Gate Recurrent Unit).

2.3.1 RNN

RNN is a neural network structure for processing sequential data. Unlike a feedforward neural network whose output depends only on the current input, the computed result of RNN also relies on the output of the previous hidden layer, which has higher performance in processing time series data, and its basic structure is shown in Fig. 2.3.

Assuming that the input is x_t at the moment t, the value of the hidden layer s_t and the output o_t can be calculated as,

$$\begin{aligned} s_t &= f(U \cdot x_t + W \cdot s_{t-1}) \\ o_t &= g(V \cdot s_t) \end{aligned} \tag{2.1}$$

where $f(\cdot)$ and $g(\cdot)$ are both activation functions. s_{t-1} represents the hidden layer value at the moment $t-1$. U is the weight matrix from the input layer to the hidden layer. V denotes the weight matrix form the hidden layer to the output layer. W suggests the weight matrix from the output value of the hidden layer at the previous moment to the input at the current moment.

2.3.2 LSTM

When training with the temporal backpropagation algorithm, the problem of gradient disappearance or gradient explosion is often encountered if the RNN network has many layers. If the gradient is too small, the update of the model parameters will become insignificant, and the model will stop learning. When the gradient is too

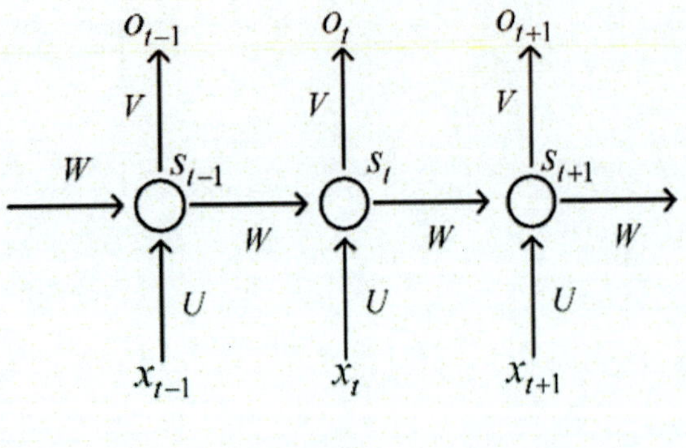

Fig. 2.3 Basic structure of RNN

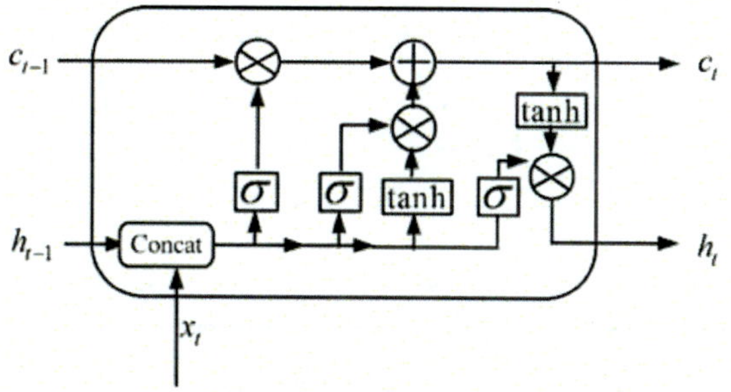

Fig. 2.4 Basic structure of LSTM

large, the model weights change more and more, and the model will make errors and fail to complete the training. To solve this problem, Hochreiter et al. proposed the LSTM to achieve long-range information memory. The structure of the LSTM is shown in Fig. 2.4.

There are three main processing steps within the LSTM: First, the forgetting gate is used to determine which step need to be remembered or not in cell states c_{t-1} from the previous. Then, the input gate is used to control the degree of memory for the current input x_t. And finally, the output gate is used to manage the current output state.

LSTM improves the memory module in the traditional RNN model by constructing a particular memory unit structure so that the information in LSTM can be updated and transmitted for a longer time, making the current prediction also available to the previous information. Therefore, LSTM has a stronger learning ability and information selection ability. As a new type of deep learning network, LSTM handles various types of temporal tasks well.

2.3.3 *GRU*

In practical applications, LSTM often requires high computational overhead due to considerable number of parameters, so researchers have proposed a simplified LSTM, GRU, which combines the forgetting and input gates of the LSTM with fewer parameters than the LSTM (Fig. 2.5).

2.4 Experiment Results

To verify the effectiveness of machine learning and deep learning-based methods in time series forecasting, this paper selects the highest daily stock price of Apple Inc. for the period from 2017-10-13 to 2022-10-11 as the dataset for model training and inference.

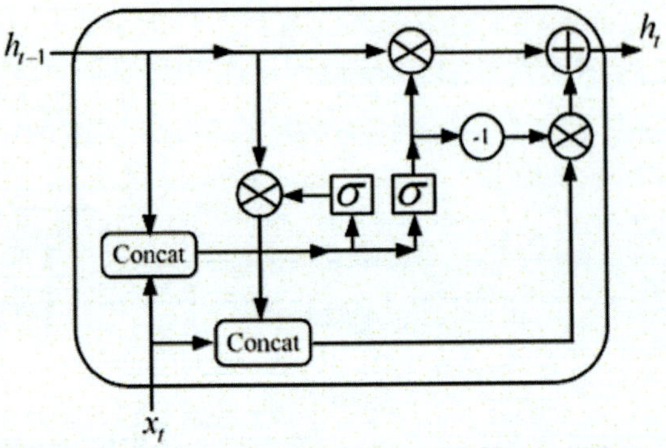

Fig. 2.5 Basic structure of GRU

To objectively evaluate the accuracy of the prediction results, quantitative indicators MAPE and SMAPE are introduced, which are defined as, respectively.

$$\begin{aligned}\mathrm{MAPE}(y, \hat{y}) &= \frac{100\%}{n}\sum_{i=1}^{n}\left|\frac{y_i - \hat{y}_i}{y_i}\right| \\ \mathrm{SMAPE}(y, \hat{y}) &= \frac{100\%}{n}\sum_{i=1}^{n}\left|\frac{y_i - \hat{y}_i}{\frac{|y_i|+|\hat{y}_i|}{2}}\right|\end{aligned} \tag{2.2}$$

where y denotes the true series data and $\hat{y}$ represents the predicted series. The value of both sequence is in the range of 0 and 1. The smaller the value calculated, the closer the predicted series is to the actual data and the higher the prediction accuracy.

2.4.1 *Machine Learning Results*

A rolling procedure is performed on the dataset to get multiple samples based on which features are extracted. The rolling, feature extraction, and feature selection are implemented using the tsfresh library concurrently. All three machine learning methods are implemented using the sklearn library, and the parameters are default values. The forecasting results of the three machine learning methods are shown in Fig. 2.6, where the red line represents the true values, while the blue line denotes the predicted ones. Intuitively, there is a better fit between the predicted and actual values using the ordinary least squares linear regression method, while the prediction seems incomplete using random forest. Table 2.2 gives a quantitative comparison of the three methods. As shown in the table, the model training time and prediction accuracy are significantly better when least squares linear regression is used than the two compared methods.

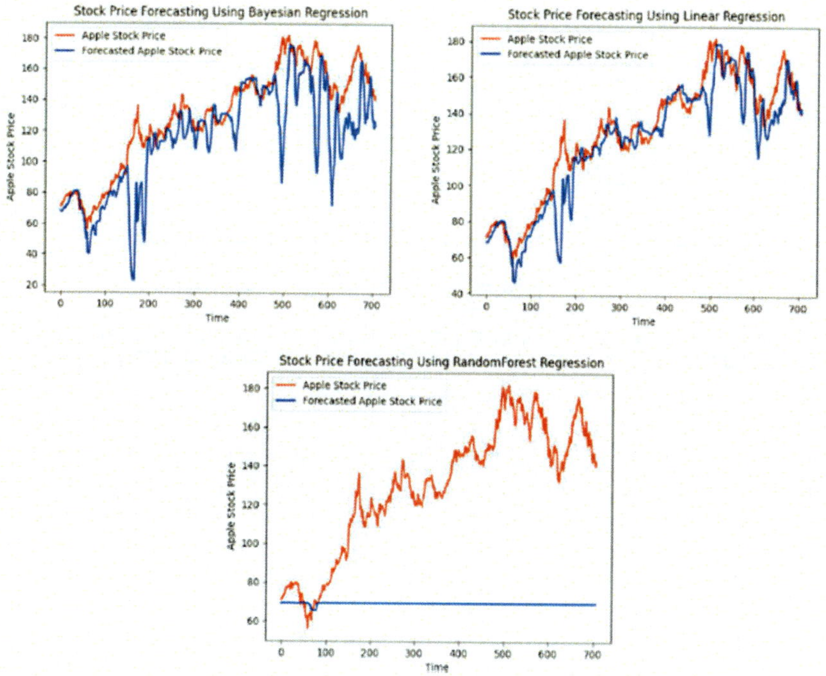

Fig. 2.6 Comparison of machine learning methods

Table 2.2 Quantitative comparison of machine learning methods

Method	Training time (s)	MAPE (%)	SMAPE (%)
Bayesian regression	0.071	12.45	14.77
Linear regression	**0.007**	**7.03**	**7.64**
Random forest regression	0.234	42.52	56.71

2.4.2 Deep Learning Results

The deep learning methods are implemented using the TensorFlow framework, and each model is trained for ten epochs using Adam optimizer. The experimental results of prediction using deep learning methods are shown in Fig. 2.7, and the predicted series data of all three manners are close to the actual data. Table 2.3 gives the quantitative comparison of the three methods. The RNN model has a more straightforward structure and fewer parameters than the GRU and LSTM. Still, its training process takes the longest time.

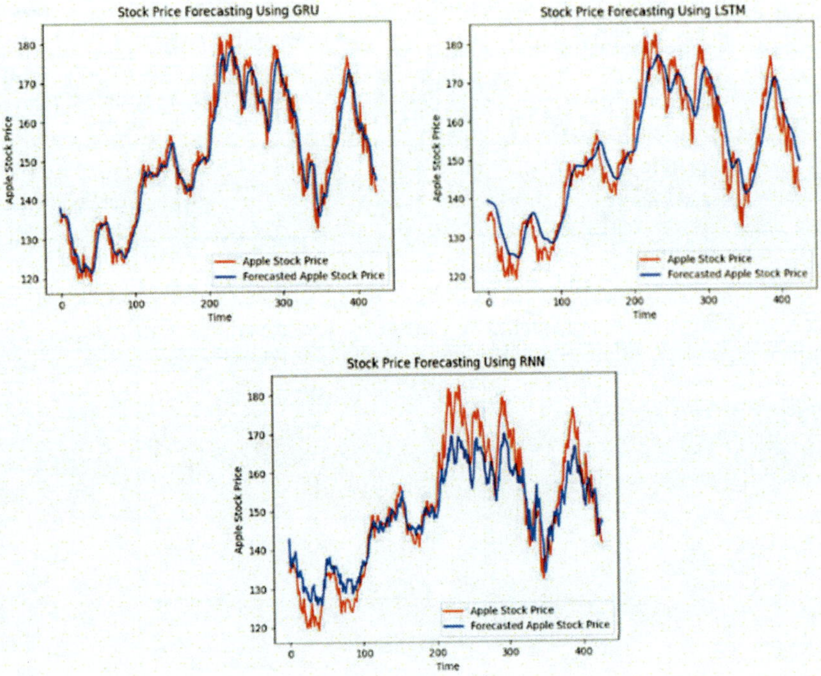

Fig. 2.7 Comparison of deep learning methods

Table 2.3 Quantitative comparison of deep learning methods

Method	Training time (s)	MAPE (%)	SMAPE (%)	Parameters
GRU	**2.739**	**1.87**	**1.87**	74,621
LSTM	2.798	2.98	2.96	98,741
RNN	3.026	3.10	3.12	**24,761**

A side-by-side comparison of machine learning and deep learning methods shows that machine learning models are faster in training but have much less predictive accuracy than deep learning methods. The deep learning approach trades off a more complex model for higher prediction accuracy.

2.5 Conclusion

In this paper, we validate the effectiveness of machine learning and deep learning methods for univariate time series forecasting problems using Apple's daily stock price data as a time series. Among them, the deep learning method achieves more accurate prediction results. It eliminates the tedious feature extraction and feature selection process required by the machine learning method at the cost of longer training time for the forecasting model. More complex combinatorial models will be considered in future research work to achieve more accurate predictions. Also, the prediction scenario will be converted from univariate time series to multivariate time series forecasting. This technique, a subtype of machine learning, alters the way we see the connection between problem solving and analytics. Deep learning allows the data to train the computer rather of teaching it how to think, creating prediction models that get better with each new batch of input data.

References

1. Duran, N., Catak, M.: Forecasting of wind speed by means of window-shifted autoregressive time series. In: IEEE 24th Signal Processing and Communication Application Conference (SIU) (2016)
2. Yasmeen, F., Sharif, M.: Forecasting electricity consumption for Pakistan. Int. J. Emerg. Technol. Adv. Eng. **4**(4), 496–503 (2014)
3. Eljazzar, M.M., Hemayed, E.E.: Enhancing electric load forecasting of ARIMA and ANN using adaptive fourier series. In: IEEE 7th Annual Computing and Communication Workshop and Conference, pp. 1–6 (2017)
4. Labouar, A., Slama, J.B.H.: Hour-ahead wind power forecast based on random forests. Renew. Energy **109**, 529–541 (2017)
5. Hu, Q., Zhang, S., Xie, Z., et al.: Noise model based v-support vector regression with its application to short-term wind speed forecasting. Neural Netw. **57**, 1–11 (2014)
6. Lin, Y., Huang, X., Chun, W.D., et al.: Early warning for extremely financial risks based on ODR-ADASYN-SVM. J. Manage. Sci. China **19**(5), 87–101 (2016). (In Chinese)
7. Huang, Q., Wei, S.: Improved quantile convolutional neural network with two-stage training for daily-ahead probabilistic forecasting of photovoltaic power. Energy Convers. Manage. **220**, 113085 (2020)
8. Zeroual, A., Harrou, F., Dairi, A., et al.: Deep learning methods for forecasting CoVID-19 time series data: a comparative study. Chaos Solitons Fract. **140**, 110121 (2020)
9. Zhang, P., Ci, B.: Deep belief network for gold price forecasting. Resour. Policy **69**, 101806 (2020)
10. Shi, J., Guo, J., Zheng, S.: Evaluation of hybrid forecasting approaches for wind speed and power generation time series. Renew. Sustain. Energy Rev. **16**(5), 3471–3480 (2012)
11. Hu, J., Wang, J., Zeng, G.: A hybrid forecasting approach applied to wind speed time series. Renew. Energy **32**(7), 82–86 (2013)
12. Darwish, A., Hassanien, A.E., Das, S.: A survey of swarm and evolutionary computing approaches for deep learning. Artif. Intell. Rev. **53**(3), 1767–1812 (2020)

Chapter 3
Research and Development of Visual Interactive Performance Test Methods and Equipment for Intelligent Cockpit

Mengya Liu, Xuan Dong, and Sheng Zhou

Abstract With the continuous growth of the intelligent configuration level of the automobile cockpit in the domestic market, the visual experience of users in the intelligent cockpit has been greatly improved. In order to ensure the safety of the driver and passengers, the safety test of the intelligent cockpit requires a quantitative test of the driver's line of sight ability. In this paper, a bionic head vision tracking method for intelligent cockpit is described, and an intelligent bionic robot is developed to simulate the driver's eye point position by adjusting the height and posture of the bionic robot. The main test is to implant a high-frame camera into the eye of the bionic robot to take and analyze the optical icon and imaging in the cockpit.

3.1 Introduction

Experimental psychologist Treicher proved through a large number of experiments that 83% of human information comes from vision. Especially in the field of intelligent cockpit, the simulation of the driver's eye ellipsoid position is the basis of cockpit optical test [1]. Meanwhile, in order to satisfy the living organism recognition algorithm in the cockpit, a pilot bionic robot is also needed in the driving position. This paper describes a kind of bionic vision robot for intelligent cockpit, which can be used for quantitative test of the driver's sight ability [2, 3]. The bionic vision robot is used to simulate the driver's height, posture and eye position. Optical devices such as binocular high-frame camera or imaging colorimeter are implanted in the eyes of the bionic robot. Optical icons and images in the cockpit were photographed and analyzed [4].

M. Liu · X. Dong (✉) · S. Zhou
Research Institute of Highway Ministry of Transport, Ministry of Transport, Beijing, China
e-mail: x.dong@rioh.cn

Key Laboratory of Operation Safety Technology on Transport Vehicles, Ministry of Transport, Beijing, China

S. Patnaik et al. (eds.), *3D Imaging—Multidimensional Signal Processing and Deep Learning*, Smart Innovation, Systems and Technologies 348,
https://doi.org/10.1007/978-981-99-1145-5_3

3.2 System Overview

The visual bionic robot is a system device to evaluate the image quality of the intelligent cockpit. The system is composed of the main body of the robot, binocular high-frame camera, built-in head, main box, software and so on, as shown in Fig. 3.1. The vision bionic robot has the ability of six-axis motion, and can realize the six-axis attitude adjustment (X axis, Y axis, Z axis, pitch, roll, heading) of the head binocular high-frame camera. Its built-in stable head can provide conditions for the high-frame camera to shoot stable and clear images. The system refers to "GB/T 11,563-1995 Automobile H-point determination Procedure" and "GB/T 36,606-2018 Ergonomic Vehicle Driver's eye position" and other standards. It can be applied to electronic rearview mirror test, HUD test, center control screen and dashboard test, multimedia rearview mirror test and other tests [2, 5].

3.2.1 *Visual Bionic Robot*

Visual bionic robot is the core part of this system. The bionic robot is composed of X axis module (X axis, X axis motor), Y axis module (Y axis, Y axis motor), Z axis module (Z axis, Z axis motor), self-stabilizing head (head pitch shaft, head roll shaft, head course shaft), prosthetic butt base and prosthetic leg, as shown in Fig. 3.2.

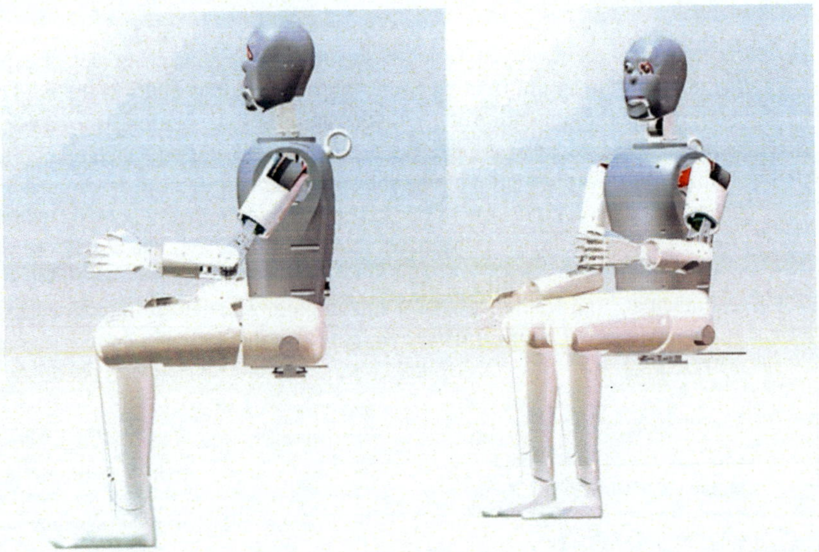

Fig. 3.1 System model of bionic robot

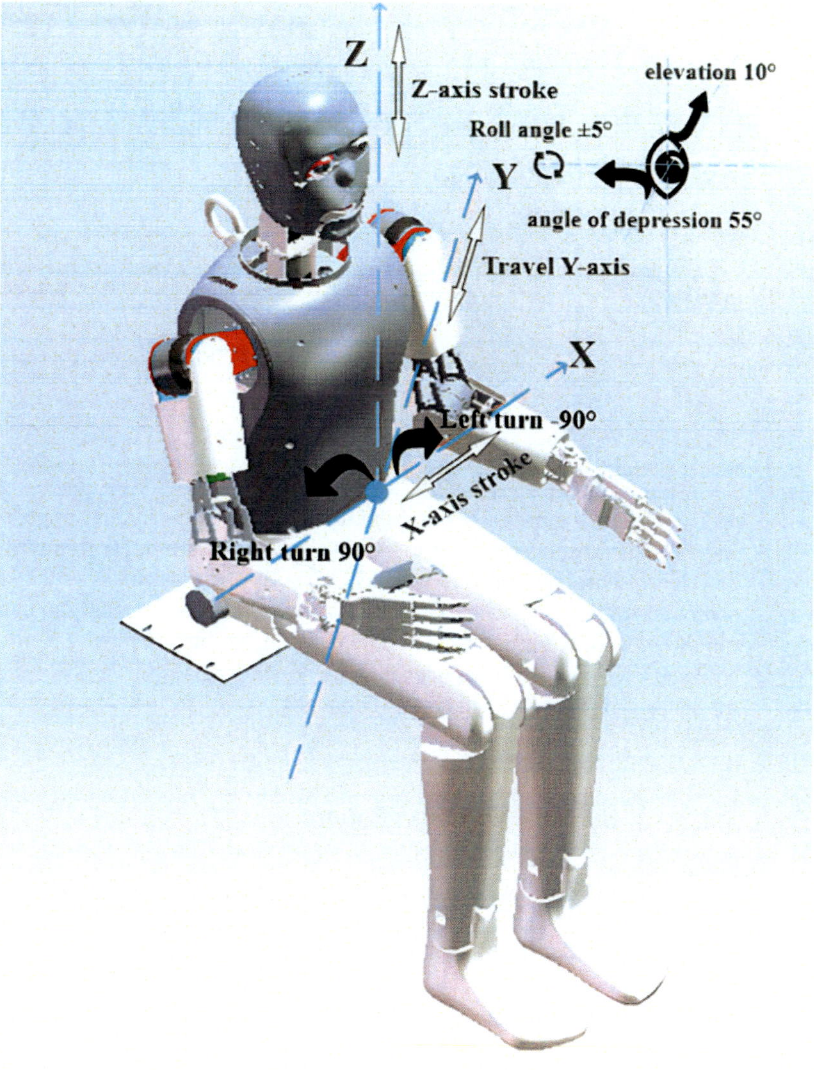

Fig. 3.2 Composition of bionic robot

X axis module to achieve X axis motion range; Y axis module to realize Y axis motion range; Z axis module to achieve Z axis motion range; The self-stabilized cradle head realizes the three-axis self-stabilization of the cradle head and provides a guarantee for the stable and clear image shooting by the imaging colorimeter. The fake butt base simulates the shape of the human butt and provides a stable base for the system. The robot can be firmly combined with the seat through the bracket.

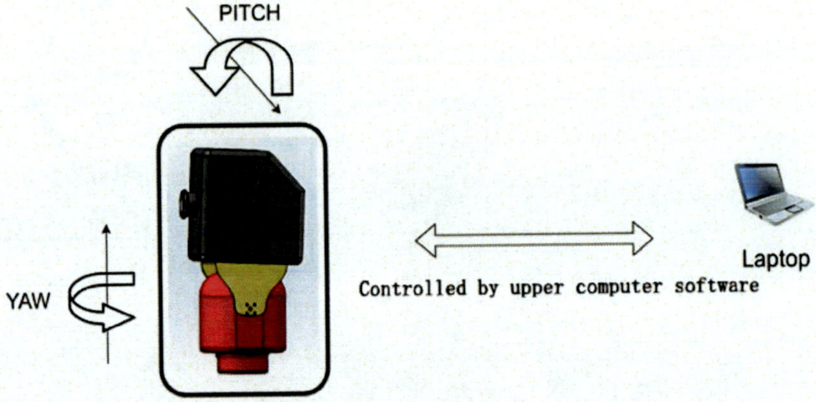

Fig. 3.3 Control schematic diagram

The motion control of bionic robot adopts the combination of upper computer software and absolute encoder to carry out precision control and movement. The function of absolute encoder can make its positioning accurate, so that the upper computer software can carry out zero calibration when initializing, and then make it run through software and serial communication, as shown in Fig. 3.3. After multi-circle absolute value calibration, the initial positioning position will not slide except for battery consumption and battery deduction.

3.2.2 Main Case of Bionic Robot

The main box integrates the motor drive of the bionic robot, which is used for power supply and control of the bionic robot bracket. The main box has 6 interfaces, which are power interface, computer connection interface, encoder interface, motor line, brake interface, limit interface and communication interface.

3.2.3 Binocular High-Frame Camera

The binocular high-frame camera is the origin of the bionic robot's eye position. The binocular camera adopts binocular parallax stereoscopic imaging technology and consists of two cameras, positioning lights and other modules to realize the calibration and reconstruction of the imaging colorimeter attitude, as shown in Fig. 3.4. Communication mode can use Modbus Tcp communication protocol, Tcp/Ip communication protocol or SD function call interface and can coordinate the work of multiple machines.

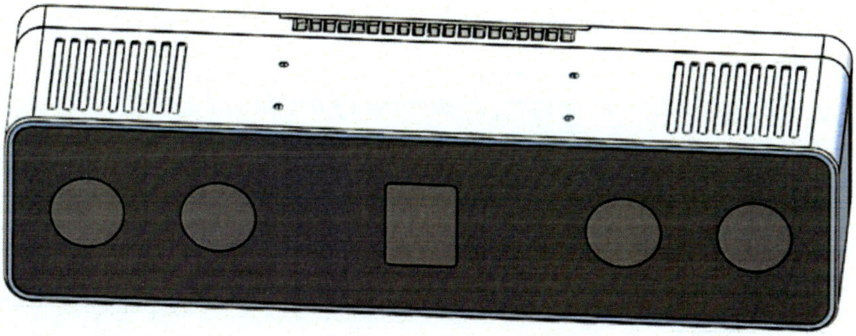

Fig. 3.4 Model of binocular camera

3.2.4 *Software*

The bionic robot test system software combined with the binocular camera to adjust the bionic robot's eye posture, as shown in Fig. 3.5. The position information and posture of X axis, Y axis and Z axis of the self-stabilizing head of the bionic robot can be set. When the angle and attitude appear slight deviation, can be fine-tuned to make the angle and attitude appear in line with the test.

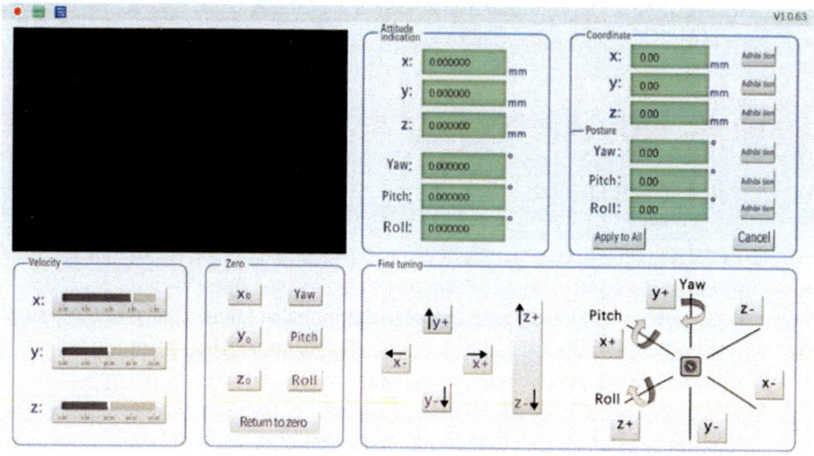

Fig. 3.5 Test system software of bionic robot

3.3 Head Visual Tracking

3.3.1 Self-Stabilizing Function of the Head

The head visual tracking function is self-stabilizing for the head. When the head is in the test, the angle deviation of X axis stroke, Y axis stroke and Z axis stroke occurs, the head can be automatically adjusted to the initial position. The self-stabilizing system design scheme is modified by high-precision common head and stabilizer. The self-stabilizing control system is composed of high-precision servo motor, encoder and self-stabilizing controller.

3.3.2 High-Precision Servo Motor

High-precision servo motor is a micro special motor used as the actuator in the self-stabilizing function of the head. Its function is to convert the electrical signal into the angular speed of the rotating shaft. The high-precision servo motor is used as the actuator in the automatic control system to convert the received electrical signal into the angular velocity output on the motor shaft. This system uses 6025 brushless motor, which is a specially designed control motor for the head, with high-precision, fast response, large torque and so on.

3.3.3 Servo Encoder

The function of servo encoder is to correct the motion deviation of high-precision servo motor, so that the motor can work strictly according to the instructions of PLC. Because of the existence of the servo encoder, the servo motor can realize the closed-loop control, and the servo system has the control accuracy far beyond that of other motors. The principle is to use the magnetic field changes to obtain the motor rotation angle data, through the controller to make the motor quickly return to the initial position. The advantages of using the encoder are: (1) debugging is simpler and faster, without a lot of heavy debugging and calibration work; (2) more power saving, endurance is five times longer than the common brushless head without encoder; (3) Only one IMU is needed, no longer the FRAME IMU; (4) Fast response, the motor instantaneous torque increases, the head is more stable, no pressure at various angles.

3.3.4 Self-Stabilizing PID Algorithm for the Cradle Head

The self-stabilizing function of the head is realized by PID algorithm. PID control is a closed-loop control system with mature technology, easy to be familiar with and master, no need to establish a mathematical model, good control effect and robustness. Its simple structure and convenient adjustment have become one of the mainstream technologies of industrial control. The control schematic figure is as follows (Fig. 3.6).

PID control is composed of three control parts: proportional regulation part, integral regulation part and differential regulation part. The control principle is: the input is output to the actuator after the operation of proportion, integration and differentiation, and a closed-loop system is formed by transmitting the operation through a negative feedback measuring element. The motor control system outputs the input signal to the motor through PID control technology, and then the feedback element carries out a new calculation on the feedback information, and then inputs it to the motor to form a closed-loop system.

The PID simulation formula is:

$$u(t) = K_p\left[e(t) + \frac{1}{Ti}\int_0^t e(t)dt + Td\frac{de(t)}{dt}\right]$$

The PID formula is discretized and the incremental PID formula is derived:

$$\begin{aligned}\Delta \mathbf{u(k)} &= \mathbf{u(k)} - \mathbf{u(k-1)} \\ &= \mathbf{k_p[e(k) - e(k-1)] + K_I e(k) + K_D[e(k) - 2e(k-1) + e(k-2)]}\end{aligned}$$

where deviation = target value − feedback value $e(k) = r(t) - y(t)$

K_p: proportional gain, adjustment parameter;

K_i: integral gain, adjustment parameters;

K_d: differential gain, adjustment parameter;

e: Deviation = target temperature − current temperature;

t: The time is now.

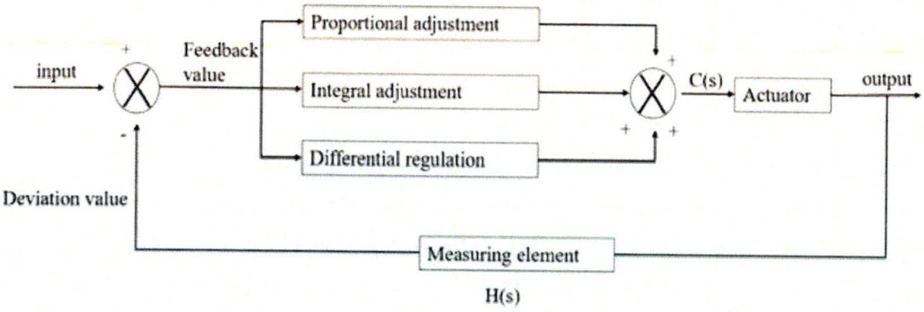

Fig. 3.6 Self-stabilizing principle model of bionic robot cradle head

Table 3.1 Precision test table of self-stabilizing function of head

No	Moving value/mm	Encoder feedback value/mm	Moving accuracy/μm
1	45	45.01	22.217
2	20	19.99	50.025
3	50	50.01	19.996
4	30	29.99	33.344
5	80	80.01	12.498

3.3.5 *Precision Test of Self-Stabilizing Function of Head*

Calibrate self-stabilizing cradle head, stable without jitter under the condition of no encoder. The motor is always on during the debugging of the encoder. The self-stabilizing function of the head, the accuracy is up to 50 μm, the specific values are shown in the table below. The above method was adopted for data recording in the test, and a total of 5 tests were recorded. The calculation method of moving accuracy was as follows: moving accuracy = (encoder feedback value − moving value)/10^5 (Table 3.1).

3.4 System Effect

The bionic vision robot is placed in the driver's seat, the main case of the bionic robot is placed in the passenger seat, and the binocular camera bracket is attached to the binocular camera on the right glass window. Firstly, the relative coordinate points of the bionic robot were measured and calibrated by combining level and ruler scale.

The position angle and attitude of X axis, Y axis and Z axis are included in the software of the bionic robot test system. The position of the bionic robot eye point is aligned with the central control screen, and the characters, colors, brightness and other data on the central control screen are obtained for analysis and evaluation, as shown in Fig. 3.7.

3.5 Conclusion

The visual bionic robot system of this system is applied to electronic rearview mirror test, HUD test, center control screen and dashboard test, multimedia rearview mirror test and other tests in the intelligent cockpit. The function and performance of the equipment can be verified by standardized tests according to the corresponding

Fig. 3.7 Adjusting the angle of the bionic robot

national standards and row standards. It is hoped that the research and development of this system can make a contribution to the standardized testing of the testing industry.

Acknowledgements This work was supported by the Central Public Research Institutes Special Basic Research Foundation (2022-9026).

References

1. Li, H.Y., Luo, J., Chen, J.B., Liu, Z.X., Xie, S.R.: Development of robot bionic eye with spherical parallel manipulator based on oculomotor control model. Przegląd Elektrotechniczny **88**(1B), 1–7 (2012)
2. Wang, X.Y., Zhang, Y., Fu, X.J., Xiang, G.S.: Design and kinematic analysis of a novel humanoid robot eye using pneumatic artificial muscles. J. Bionic Eng. **5**(3), 264–270 (2008)
3. Villgrattner, T., Ulbrich, H.: Design and control of a compact high-dynamic camera-orientation system. IEEE/ASME Trans. Mechatron. **16**(2), 221–231 (2011)
4. Gu, J., Meng, M., Cook, A., Faulkner, M.G.: A study on natural movement of artificial eye implant. Robot. Auton. Syst. **32**(2), 153–161 (2000)
5. Bertozzi, M., Broggi, A., Fasciol, A., Nichele, S.: Stereo vision-based vehicle detection. In: Proceedings of the IEEE Intelligent Vehicles Symposium 2000 (Cat. No. 00TH8511), pp. 39–44. IEEE (2000)

Chapter 4
Design and Validation of Automated Inspection System Based on 3D Laser Scanning of Rocket Segments

Jigang Chen, Shunjian Ye, Lu Jin, Jinhua Chen, and Zhiyong Mao

Abstract To address the problem of low degree of automation and low efficiency of geometry measurement of large size rocket segments, an automated inspection system based on 3D laser scanning is designed. The 3D laser scanner is the main measurement equipment, supplemented by the lifting mechanism, etc. The geometry measurement of the module segment is realized by the segment scanning and splicing method. The dimensional error and form error of the measurement system reach 0.2 mm. In order to verify the actual measurement accuracy of the automated inspection system, a field accuracy verification test was designed, and the verification results showed that the measurement index of the system met the design requirements.

4.1 Introduction

The height of rocket segments usually ranges from 1 to 4 m, and the diameter ranges from 2 to 4 m, so the traditional measurement equipment can no longer meet the requirements for the measurement of the shape and size of rocket segments [1, 2]. At present, the theodolite measuring system used to inspect the rocket module needs to stick the target manually and arrange the measuring points, the automation is

J. Chen · Z. Mao (✉)
Shanghai Precision Metrology and Testing Research Institute, Shanghai, China
e-mail: maozhiyong99@163.com

J. Chen
e-mail: cjg20109@126.com

S. Ye · L. Jin · J. Chen
Shanghai Spaceflight Precision Machinery Institute, Shanghai, China
e-mail: ysjzj85@126.com

L. Jin
e-mail: bobking119@163.com

J. Chen
e-mail: chen13636505484@163.com

S. Patnaik et al. (eds.), *3D Imaging—Multidimensional Signal Processing and Deep Learning*, Smart Innovation, Systems and Technologies 348,
https://doi.org/10.1007/978-981-99-1145-5_4

in a low level. The measuring range is restricted, and several personnel need to cooperate at the same time when measuring, so the production tasks and higher accuracy requirements, it is necessary to improve and research the measurement methods of large space dimensions, in order to improve the degree of measurement automation, increase work efficiency and meet the needs of scientific research and production.

In recent years, 3D laser scanning technology has been widely used in mapping, reverse engineering and other fields [3]. The 3D laser scanning technology uses non-contact measurement, which has the characteristics of high accuracy of point measurement, high density of spatial point acquisition, high speed, and the possibility of flexible measurement, etc. The laser is used to scan and obtain 3D coordinates and morphological information of different points, and according to this information data, a 3D model with high accuracy can be obtained [4]. A large number of domestic and foreign studies have applied laser scanners to the measurement of various types of parts [5, 6].

In this paper, we design an automated inspection system based on 3D laser scanning of rocket segments to address the limitations of low efficiency, low automation, and excessive personnel intervention by using the theodolite measuring system. The laser scanner as the main measuring device, a fixed lifting mechanism as the supporting transfer station mechanism, and scanning in segments according to the height of the measured segment. The final point cloud stitching method is used to achieve the geometry measurement of the whole module. The accuracy verification experiment was designed to verify the reliability and measurement accuracy of the designed automated inspection system.

4.2 3D Scanning Measurement Principle

3D laser scanning is a 3D visual sensing technology that uses the laser ranging principle to obtain the spatial information of the measured target surface at high speed. According to the laser ranging principle, it can be divided into two categories: phase type and pulse type. Phase laser have short ranging distance and high measurement precision, while pulsed laser have long ranging distance and low measurement precision. Considering the shape of the measured structural parts and inspection accuracy requirements, this paper uses phase laser measurement method.

Phase laser measurement is to modulate the amplitude of the laser beam by the frequency of the radio band, and determine the phase delay generated by the modulated light propagating, and then convert the distance corresponding to this phase delay according to the wavelength of the modulated light. In other words, the laser transmitter emits a modulated continuous wave to the target to be measured, and the continuous wave is returned after reaching the target to be measured. The distance between the measuring device and the target to be measured is calculated based on the wavelength of the modulated wave and the phase delay between the transmitted and received waves. As shown in Fig. 4.1.

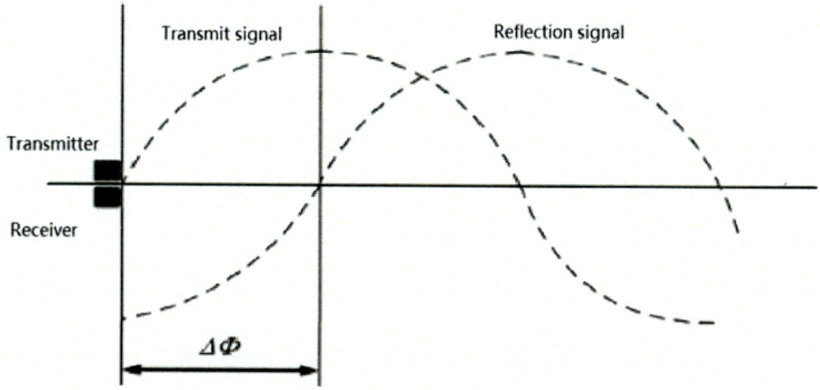

Fig. 4.1 Schematic diagram of phase laser ranging

The distance between the phase laser ranging equipment and the target to be measured is:

$$D = \frac{\lambda}{4\pi} \cdot \Delta\Phi \quad (4.1)$$

where λ is the wavelength of the modulated wave, and $\Delta\Phi$ is the phase difference between the transmitted wave and the received wave

Accurate 3D laser scanning point cloud data are obtained based on 3D laser scanning equipment, and each data point is 3D coordinate data. After processing and analyzing the 3D point cloud data of the measured object, the measured values of different geometric sizes are calculated by data fitting algorithms.

Automatic Inspection System Construction.

The automated inspection system mainly contains 3D laser scanning equipment, lifting platform system, control acquisition terminal, verification standard, and inspection software. The composition of the inspection system is shown in Fig. 4.2.

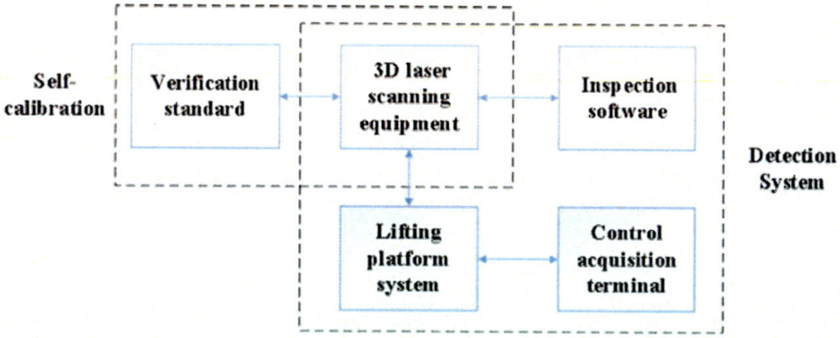

Fig. 4.2 Schematic diagram of the automated inspection system

4.2.1 Three-Dimensional Laser Scanning Equipment

The 3D laser scanning equipment is based on the principle of phase laser ranging. By recording the 3D coordinates, reflectivity, and texture of a large number of dense points on the surface of the object under test, the 3D model of the target under test and various 3D data such as lines, surfaces, and bodies can be quickly reconstructed. At the same time with the use of the camera system, the shape of the target under test can be quickly identified.

4.2.2 Lifting Platform System

The lifting platform system is composed of lifting module and control system. The lifting platform is fixed on the ground through the ground bolts, and the lifting module adopts electric interlocking large spiral lifting mechanism; lifting movement repeatability: $\leq$10 mm; lifting stability: shaking amplitude $\leq 8'$. The control system is composed of upper computer system and network communication system, which can realize the operation status monitoring and remote control. As shown in Fig. 4.3.

4.2.3 Checking Standard Device

The checking standard device includes the hole distance standard and the ball bar standard. The hole distance standard is a piece of physical standard with hole center distance of 1000 mm, and there are two small holes of Φ10 mm on the apparatus, which is used to verify the hole measuring ability of the laser scanning equipment. The ball bar standard is a piece of physical standard with ball center distance of 800 mm, which is used to verify the detection ability of the laser scanning equipment.

4.2.4 Measurement Software Design

The measurement software is used to realize the control of the main measuring equipment (including start-up, adaptive calibration, and measurement), the control of the fixed lifting platform system, the stitching calculation of the measurement data, and the measurement result report according to the demand. The measurement software has certain expandability, which providing point cloud data analysis algorithm and control interface of fixed lifting mechanism, such as API function, debugging window.

The software system is divided into four layers of architecture: UI layer, application layer, support layer, and system layer.

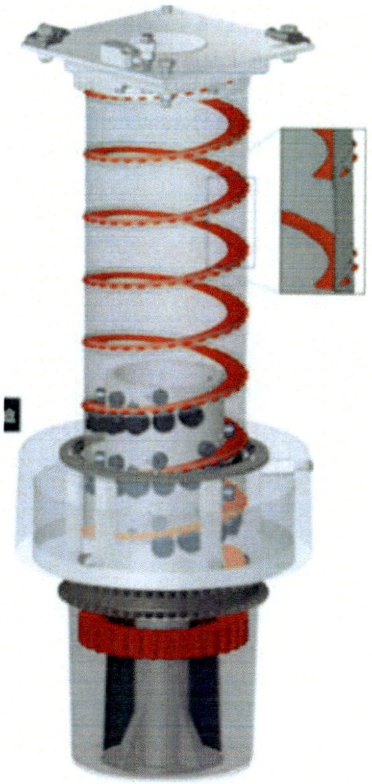

Fig. 4.3 Schematic diagram of the lifting platform system

The UI layer is the interface that directly interacts with the user and is the visualization package of the main front-end workbench. The application layer is the core part of the software system, responsible for completing the functions of equipment monitoring and control, data processing and analysis, report generation, personnel management, logging, etc. The support layer is the middleware that connects the system layer and the application layer, responsible for providing dynamic link libraries for software development, communication interfaces, software development modules, etc. The system layer is the foundation of the software system and provides the environment for the operation of the software system.

4.3 Measurement System Accuracy Verification

In order to fully verify the feasibility and accuracy of the measurement system designed in this paper, considering the actual working conditions such as large size and complex geometric elements of the measured rocket segment, the following two aspects need to be verified.

(1) Length splicing accuracy verification, to ensure that the distance measurement accuracy in large sizes meets the requirements.
(2) Geometric element detection verification, to ensure that the accuracy of detecting small holes in the rocket module meets the requirements.

4.3.1 Verification of Length Splicing Accuracy

The maximum size of the measured module is about 4 m, and several geometric elements are distributed in different positions and quadrants of the module, so the laser scanner is required to perform multi-station splicing test to complete the overall measurement. The laser tracker is used to measure 9 measurement points on the three-dimensional space. The distance of each measurement point of the tracker is used as the distance measurement standard value.

In order to ensure the accuracy and validity of the standard value of the distance measurement, the tracker was repeatedly measured twice consecutively, and the results of the two measurements were fitted to each other for verification to ensure that there was no gross error in the standard value, and the four distances of the tracker (point 1–point 7), (point 4–point 7), (point 4–point 8), and (point 5–point 8) were taken as the standard value, and the standard quantity values are shown in Table 4.1.

The laser scanner performs 3 turnaround measurements, as shown in Fig. 4.5. The coordinates of 9 measurement points in the scanning space are scanned, and the coordinates of each point are stitched together using the 4 common points closest to the equipment as transfer stations. The upper computer took the distance from station one measurement point 1 to station two measurement point 7, station one measurement point 8 to station three measurement point 5, station three measurement point 4 to station two measurement point 8, and station three measurement point 4 to station two measurement point 7 as the actual measurement results, and the positions are shown in Fig. 4.4.

The distance measurement results are shown in Fig. 4.5.

4.3.2 Verification of Geometric Element Detection Accuracy

The measured cabin section has a variety of geometric elements, among which the measurement of φ10 mm small holes is particularly important. The measurement fit

Table 4.1 Standard value of distance measurement at each point

Measuring equipment	Standard value of distance (mm)			
	Point 1–7	Point 4–7	Point 4–8	Point 5–8
Laser tracker	9879.34	10,090.56	11,276.43	11,242.00

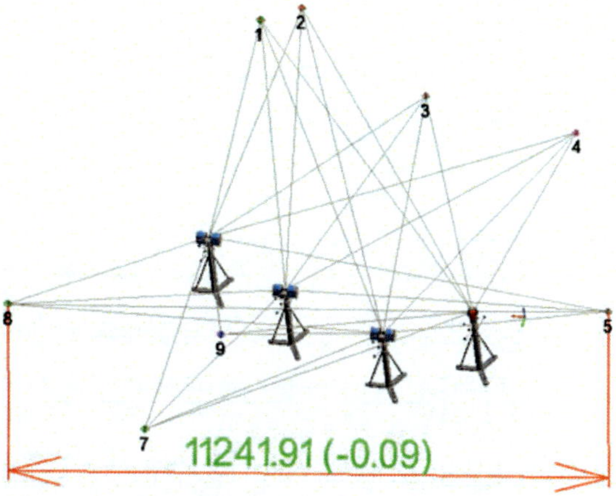

Fig. 4.4 Laser scanner transfer station location map

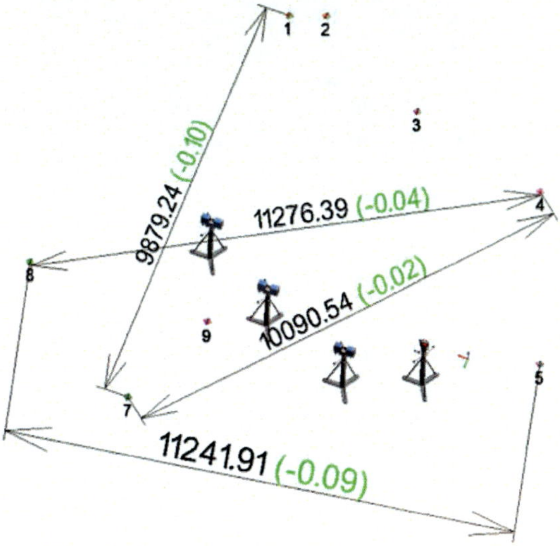

Fig. 4.5 Laser scanner measurement results

of multiple small holes is directly related to the accuracy of key inspection items such as cabin roundness and perpendicularity. Therefore, this verification is mainly for the detection accuracy of φ10 mm small holes. The verification standard is an aluminum-magnesium alloy long rod, with four φ10 mm holes on each side of the rod to simulate the physical state. The aluminum-magnesium alloy long rod is tested on a high-precision coordinate measuring machine to obtain the standard value, and the site is shown in Fig. 4.6.

The whole verification will test the laser scanner in four states: front view, top view, elevation view, and 45° view. For the front view measurement verification, the standard rod will be placed directly in front of the laser scanner, the hole under test is facing the laser scanner (the axis of the hole is horizontal), the standard

Fig. 4.6 Hole center distance detection verification

rod is horizontal, and the placement distance is 1.5 and 4 m, respectively; for the elevation view measurement verification, the standard rod will be placed above the offside front of the laser scanner, the axis of the hole under test is vertical, and the placement distance is 1.8 m and the height is about 0.8 m, respectively; for the top view measurement verification. The standard rod is placed below the offset front of the laser scanner, the axis of the measured hole is vertical, the placement distance is 1.8 m, and the height is about 0.8 m, respectively; the measurement is verified by 45° view. The standard rod was placed directly in front of the laser scanner, placed diagonally at 45°view, and the measured hole was facing the laser scanner (the axis of the hole was in a horizontal state), and the placement distances were 1.5 and 4 m, respectively. The instrument obtained the distances from hole 1 to hole 2 and hole 3 to hole 4, respectively, and the test results are shown in Table 4.2.

4.4 Conclusion

Through experimental verification, the detection system designed in this paper can realize automatic measurement of the cabin segment within 10 m measurement range, and can realize the functions of measurement management, data processing, data analysis and result generation, realizing the high integration of measurement

Table 4.2 Test results

Test state	Test distance (m)	Distance standard value (mm)		Measurement results (mm)		Error of the value (mm)	
		Hole1 → 2	Hole3 → 4	Hole1 → 2	Hole3 → 4	Hole1 → 2	Hole3 → 4
Front view	1.5	55.263	57.566	55.26	57.62	0.00	0.05
	4			55.22	57.55	−0.04	−0.02
Elevation	1.8			55.39	57.43	0.13	−0.14
Overhead	1.8			55.27	57.63	0.01	0.06
45°	1.5			55.33	57.67	0.07	0.10
	4.0			55.22	57.37	−0.04	−0.20

and control as well as the unified management of measurement data. According to Fig. 4.4, 4.5 and 4.6 and Tables 4.1 and 4.2 test results, the automatic inspection system designed in the paper achieves 0.2 mm accuracy in dimensional measurement within 10 m measurement range and 0.2 mm accuracy in geometric element detection within 4 m measurement range. The measurement accuracy is better than that of laser tracker in paper [7], and that of Photogrammetry in paper [8].

4.5 Discussion

3D laser scanning technology belongs to one of the very advanced technologies in the field of measurement, which can play an important role in the field of large size measurement and has good application prospects in engineering measurement. It can be used in the assembly and manufacturing of aviation and aerospace products and other scenes for installation guidance of assembly and measurement of the shape of structural parts. With the vigorous development of the highly sophisticated equipment manufacturing, 3D laser scanning technology will have better use in industrial measurement.

References

1. Lv, J.J., Bai, J.B., Hou, D.X., Cao, J., Zhang, P.G.: Research on riveting and assembling scheme of carrier with large diameter cabin segment. Missiles Space Veh. **380**, 95–99 (2021)
2. Yang, Z.B., Li, L.Z., Zhang, P., Zhang, R.: 3D digital assembly process for rocket body cabin. Meas. Control Technol. Aerosp. Mater. Technol. **4**, 85–91 (2016)
3. Lv, C.H., Chen, X.P., Zhang, D.M.: Urban 3D modeling method based on 3D laser scanning technology. Sci. Technol. Eng. **10**, 166–170 (2012)
4. Cai, R.B., Pan, G.R.: A new method of multi-view cloud stitching for 3D laser scanning". J. Tongji Univ. (Nat. Sci. Ed.) **7**, 913–918 (2006)
5. Zhou, S., Xu, J., Tao, L., et al.: Measurement system of inclined circular surface diameter based on 3D laser scanning. Photoelectron Laser **28**, 630–638 (2017)

6. Deshmukh, K., Rickli, J.L., Djuric, A.: Kinematic modeling of an automated laser line point cloud scanning system. Procedia Manuf. 5, 1075–1091 (2016)
7. Zhu, Q., Yu, H.C., Chen, J.G., Bao, X.F.: Research on large size measurement method of satellite cylinder based on laser tracker. Meas. Technol. **36**, 33–36 (2016)
8. Hou, Q.Q., Gao, Z.P.: Research on calibration method of indoor large size space coordinate measuring and positioning system based on photogrammetry. Meas. Control Technol. **37**, 266–268 (2017)

Chapter 5
Research and Implementation of Electric Equipment Connectivity Data Analysis Model Based on Graph Database

Junfeng Qiao, Lin Peng, Aihua Zhou, Lipeng Zhu, Pei Yang, and Sen Pan

Abstract Effective description of electric data resources is an important basis for electric data modeling. With the access of large-scale distributed new energy, electric allocation data resources are more complex and changeable, and existing data description methods are difficult to accurately describe. Aiming at the above problems, the research on hierarchical description method of electric distribution and distributed new energy data resources based on complex network is carried out. Based on the complex network theory, the hierarchical description method of data resources is formed by comprehensively describing the external characteristics, internal elements, organizational structure, and relationships between data resources. The holographic modeling technology of electric distribution and distributed new energy data resources based on graphic computing is studied. Data modeling is an important basis for data fusion and sharing. The production, operation, control, measurement, and other business links in the field of electric distribution and distributed new energy maintain a certain degree of independence from each other, and there is an "information island" problem in the data resources of each link. In view of the above problems, research on holographic modeling technology of electric distribution and distributed new energy data resources based on graphic computing is carried out. Based on graphic computing technology, equipment is taken as the core, covering resources, assets, measurement, topology, graphics, and other all-round information, to build a holographic data model of electric distribution and distributed new energy data resources, and support the deepening integration and sharing of electric distribution data.

J. Qiao (✉) · L. Peng · A. Zhou · L. Zhu · P. Yang · S. Pan
State Grid Key Laboratory of Information & Network Security, State Grid Smart Grid Research Institute Co. Ltd, Nanjing 210014, China
e-mail: 1318558905@qq.com

S. Patnaik et al. (eds.), *3D Imaging—Multidimensional Signal Processing and Deep Learning*, Smart Innovation, Systems and Technologies 348,
https://doi.org/10.1007/978-981-99-1145-5_5

5.1 Introduction

Due to the influence of external factors such as light intensity and wind speed on the output of distributed electric generation [1], the output electric of the electric supply has certain uncertainty. When a large number of distributed electric generators are connected to the traditional distribution network [2], the structure and operation state of the distribution network are changed, resulting in two-way flow of electric flow in the system, random changes in direction and size [3]. The network loss and voltage of lines and systems are also affected by this randomness, which makes it more difficult to build the distribution system model. On the other hand, it can effectively improve the reliability index of the load points in the distribution network and the entire distribution system, and reduce the probability of load loss caused by the failure of the distribution network by reasonably selecting the location of the distributed generation to form a model island. Therefore, the distribution network modeling method proposed in this paper must take into account the instability of distributed generation, and fully consider the impact of distributed generation on existing distribution networks.

5.2 Related Work

The basic idea of multi-dimensional data modeling for the current integrated system of electric grid regulation is to use the classic star mode and snowflake mode to quickly and flexibly design multi-dimensional scenarios of electric grid regulation according to the requirements of analysts, then carry out rich dimensional definitions, and finally define the fact table of key indicators of relevant regulation scenarios and form a well-structured multi-dimensional data model of electric grid regulation. Electric big data modeling based on topology data, establish equipment mapping table according to electric network topology, orderly integrate the scattered and isolated massive data in the current electric grid, and propose electric data clustering analysis method based on graph theory using graph theory algorithm encapsulated by graph database itself; Finally, based on the graph database, two specific examples are analyzed to test the information retrieval performance and data clustering analysis function of Neo4j database. Based on the spatiotemporal data modeling of "grid one diagram", based on the graph database and graph computing technology, and taking advantage of the structural consistency, the naturalness of expression, and the intuitiveness of display between the graph data structure and the actual electric grid [4]. By integrating the overall topology of the electric grid, including the topology structure within all links of electric generation, transmission, transformation, distribution, and use, and the grid spatial data of their connection relationships, Open up the topological connection between electric grids of different voltage levels to form a "electric grid topology map" [5].

5.3 Research on the Method and Algorithm of Electric Data Modeling

5.3.1 Electric Data

Electric data refers to the data that exceeds the processing capacity of traditional database systems. It has the characteristics of "5V+1C": volume, variety, speed, value, flexibility, and complexity [6]. Data scale: traditional electric data processing does not process enough data, and cannot find the potential value of many electric data. With the improvement of data volume and data processing ability in the data age, it is possible to mine more electric business value from a large number of data; Diversity, electric data includes structured data, unstructured data, and semi-structured data [7]. It is fast and timely, including fast transmission and processing speed. It requires time sensitive and decision-making analysis, and can grasp the information of important events at the first time; Value, electricity, like money and gold, has become a new economic asset; Flexible, analysis and processing models must quickly adapt to new business requirements; Complex, new methods are needed to meet the requirements of unified access and real-time data processing of heterogeneous electric data. The key to the application of electric data is not "more" and "data". Its core value is to treat data as the same electric core asset as people and property, so that assets can create value [8].

5.3.2 Electric Data Modeling

In electric modeling, graph theory provides a theoretical basis for selecting independent and complete variables of the network, and also lays a foundation for introducing electric modeling technology and its application [9]. Graph theory and graph database application technology will be used in computer aided network analysis and electric network analysis [10]. When considering the connection relationship between components in the electric grid, each component in the network can be represented by a vertex, and the connection between components can be represented by a line segment, called an edge, so as to form an electric grid diagram. The elements of a graph are called primitives, and edges and vertices are the most basic primitives. Electric grid equipment mainly includes busbar, transformer, circuit breaker, etc. electric information communication equipment mainly includes server, storage, network, etc. For devices, the graph can be described as a vertex, which is described as "node+attribute"; The "terminal" and "terminal" between two devices are connected to form an edge, which is described as "edge+attribute".

The research of this topic describes the primary equipment and the connection relationship between the equipment with the graph computing model. A primary equipment in the electric grid is abstracted as a node, which is endowed with attribute information. The connection relationship between the equipment is abstracted as an

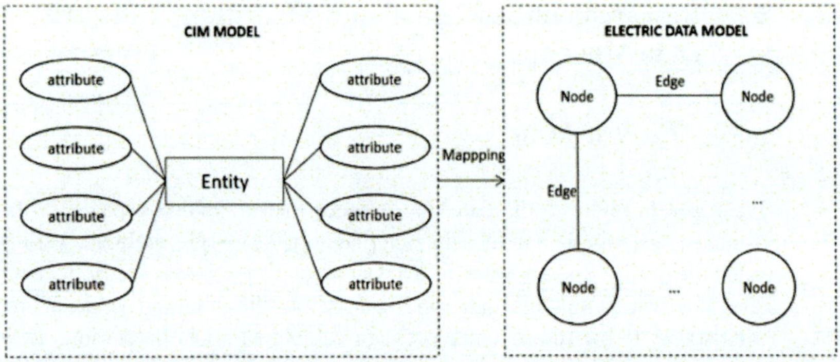

Fig. 5.1 CIM model mapped to electric data model

edge, which is also endowed with attribute information. It describes the attribute information of the connection relationship. When devices are directly connected, there must be an edge between the corresponding nodes of the equipment. On the contrary, no edges are built between these nodes. In this way, the entity attribute mode in the Common Information Model (CIM) corresponds to the node edge mode in the graph computing model, the relevant data of each device corresponds to an entity and attribute in the CIM model, the relevant data of each device in the graph computing model is a node and its attribute, and the description of the connection relationship between devices corresponds to an attribute of the entity in the CIM model, that is, attributes are used to store the description information of the connection relationship, while in the graph computing model, the connection relationship between devices is represented in the form of edges and edge attributes.

In Fig. 5.1, the entity attributes of the CIM model correspond to the points and attributes in the graph computing model, realizing the mapping between them. The point, edge, and attribute modes of the graph computing model can completely inherit the representation mode of the equipment and the relationship between the equipment in the grid CIM model.

5.4 Implementation of Electric Data Model Based on Graph Database

In accordance with the CIM model standard of the power grid, a large number of power grid physical devices are abstracted and their complex topological connection relationships are reconstructed. According to the figure, "node"+"attribute", "edge"+"attribute", "node" and "node" are uniformly described through the data description method of "edge" pointing to association, as shown in Fig. 5.2.

For the primary equipment of the electric power grid, each equipment has an equipment nameplate and other relevant information, including: equipment number,

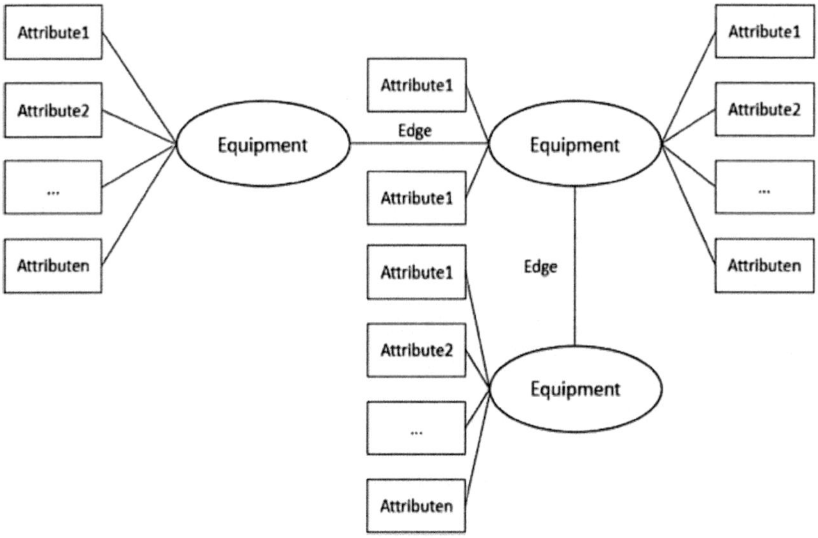

Fig. 5.2 Design of electric data model

equipment name, factory date, equipment status, detection records, and other information. For example, for the switch gear, there are "closed", "disconnected", and other states, and there is a connection relationship between the primary equipment, which is generally connected through lines. As shown in the figure above, the connection relationship and status between such devices and devices can be described comprehensively and accurately by using the elements such as points and edges in the graph computing model. Each device corresponds to a vertex or node in the graph computing model. Each vertex has one or more attributes, which can be used to describe the feature information of the vertex, the line connection relationship between devices can be represented by the edge in the graph computing model. Similarly, one or more attributes are used to describe the feature information of the edge (i.e., the line). The connection edges between vertices can be directional, directional, or undirected, and undirected, which is completely consistent with the current direction characteristics between primary devices. At the same time, the scalability and efficiency of the graph computing model can meet the unified description and representation requirements of massive, complex, and cross business grid topology, providing a model basis for subsequent topology analysis.

5.4.1 Electric Data Relation Processing

According to the coupling degree between business links and the characteristics of topology data of each power grid business link, the association rule algorithm

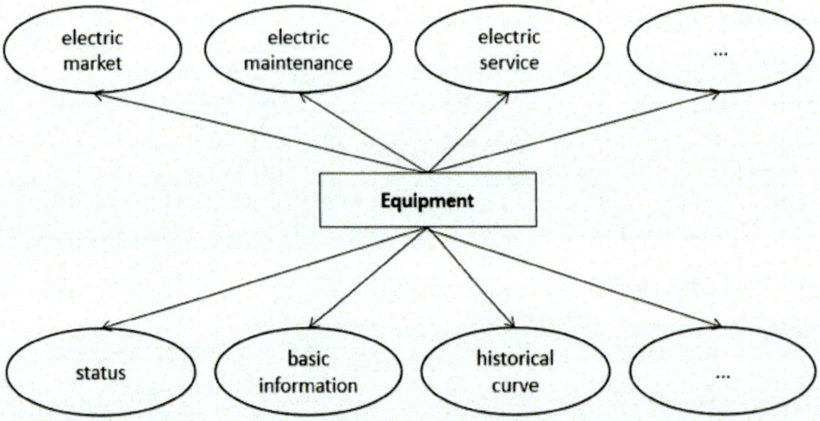

Fig. 5.3 Equipment relationships with electric application

of big data is used to fully exploit the association between power grid topology data. Through the automatic reconstruction of the relationship between power grid topology entities, the automatic association of topology data across business links and oriented to the panoramic power grid is realized.

As is shown in Fig. 5.3, in different power grid business links, there are corresponding power grid topology, and the electric topology between each business link have certain relevance. Electric data association rule mining refers to mining some possible association relationship between data from the electric data level, and the electric topology between business links must also have association relationship at the topology data level.

First, sort out the electric topology data of each business link, use big data association rule algorithm to mine the frequent patterns of topology data for each business link, find out the topology subsets with high similarity between the topologies of business links, and then judge the business rules to find out the topology intersection between two business links. The topology intersection is the associated part of the topology of two business links, so as to realize the topological connection of two business links.

5.4.2 Implementation and Construction of Power Data Model

In this paper, peak valley difference, minimum load rate, load rate, maximum load rate, power factor, power supply reliability, voltage qualification rate, three-phase imbalance rate, etc. are selected as the key parameters for the construction of the power data model. Then, dynamic weights are defined for each parameter, and the method based on variance contribution degree is used to calculate the comprehensive score of the model.

$$P_i = \sum_{j=1}^{8} C_j X_{ij} \tag{5.1}$$

Parameter C_j refers to voltage qualification rate, X_{ij} refers to the ratio of connection between device i and device j.

X1. Voltage qualification rate: total time of voltage at monitoring point within the qualified range/effective monitoring duration.

X2. Peak valley difference: peak valley difference/maximum load (report period).

X3. Minimum load rate: minimum load/maximum load (report period).

X4. Load rate: average load/maximum load (report period).

X5. Power supply reliability: average outage duration (report period) – num (voltage = 0)/96.

X6. Three phase unbalance rate: (A/B/C maximum current – A/B/C minimum current)/A/B/C maximum current.

X7. Power factor: avg = (active)/(avg*(active)^2 + avg* (reactive)^2)$^{0.5}$.

X8. Maximum load rate of distribution transformer: maximum load rate – maximum active load/rated capacity.

The input parameters of the initialization model include MapID (indicating the grid GIS map), ObjectIds (indicating the equipment), VersionNo (indicating the version and model), and the returned results include Result (indicating the results or failures, and giving the reasons) and Obejctcount (the total number of analyzed equipment nodes).

Analysis logic: according to the input parameters in the version or base network, the starting equipment is identified as electrified, with the starting equipment as the center,

(1) If the starting equipment is a transformer equipment, judge the downstream analysis of the voltage level in the current direction. As long as the topology is electrically connected, the analyzed equipment is marked with the status of electrification and the topology parent equipment of the analyzed equipment is identified as the upstream.
(2) If it is not a transformer type equipment, the breadth analysis shall be carried out. As long as the topology is electrically connected, the equipment passing through the analysis shall be identified as live and the topology parent equipment of the equipment analyzed shall be identified as upstream. When the transformer type equipment is found to have a higher voltage level, the analysis of this topology branch shall be stopped.

As is shown in Fig. 5.4, the function of the electrification initialization model is to take the starting equipment as the power point, analyze along the direction of the voltage level to the downstream of the current, and identify the passing equipment as electrified. The realization of this business logic is mainly the graph traversal search process of condition control. The key lies in how to divide the power grid into different sub graphs based on voltage level and conduct hierarchical analysis, and carry out corresponding analysis on different power sources (transformers, buses).

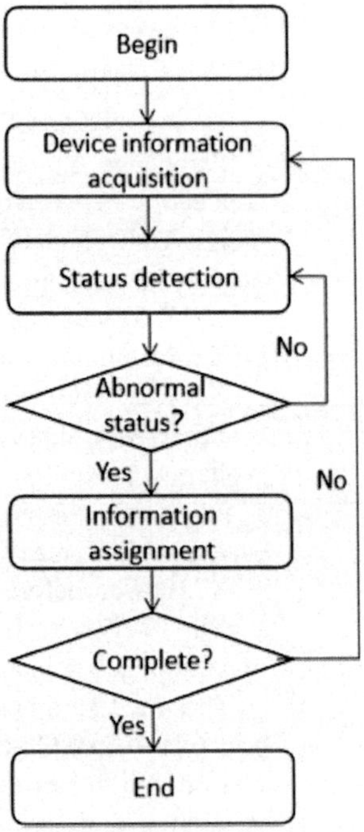

Fig. 5.4 Electric data model procession

The difficulty lies in the selection of control conditions, and how to quickly update the charged state and current direction of equipment. In R&D, iterative edge search is used to traverse the graph, and query conditions are set to quickly and accurately find and return the downstream equipment that meets the conditions. The current direction is marked by creating a new directed edge, and the topology parent device is identified by creating a reverse edge.

5.4.3 Electric Equipment Connectivity Analysis Based on Power Grid Data Model

This part is mainly used to input parameters of equipment connectivity analysis include Map_ID (indicating grid GIS diagram), ObjectIds1 (the first starting equipment node), ObjectIds2 (the second starting equipment node), is_status (indicating electrical topology or general topology analysis, 0 is general topology analysis,) Version_No (indicating version and model), the returned results include Result (indicating the result or failure, and giving the reason), object_count (the total number of analyzed device nodes), and object_Id (the ordered set of device nodes obtained through analysis).

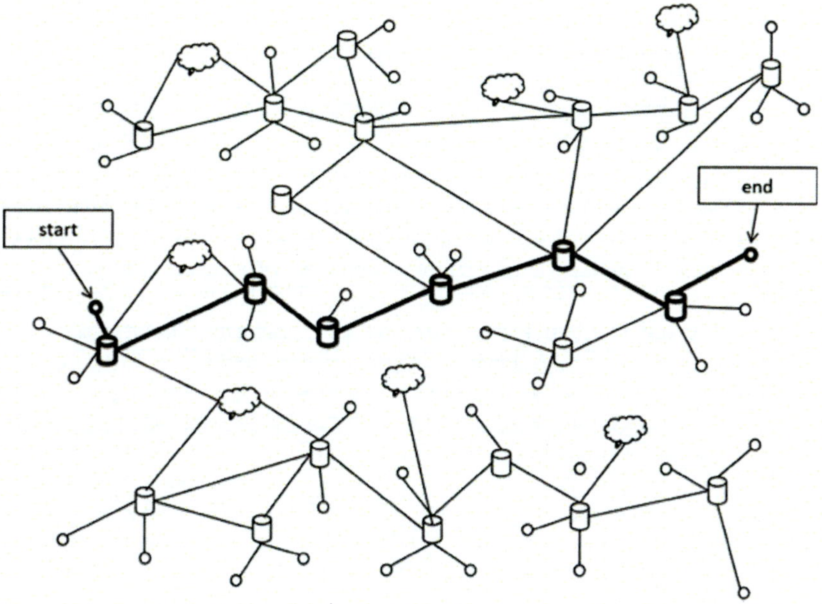

Fig. 5.5 Devices connectivity analysis

The process is mainly based on Version_No in the topology network of the base version or version: the two starting device nodes ObjectId1 and ObjectId2, which are input parameters, respectively, perform a breadth general topology analysis or a breadth electrical topology analysis according to the is_status parameter value. When a device is analyzed together, it indicates that the two devices are topologically connected and orderly form a path between the two devices, and output the path results to the corresponding output parameters, otherwise, it will not work.

In the Fig. 5.5, the two devices connectivity analysis function performs general topology analysis or electrical topology analysis on two starting devices. When there is a path, the query result will be returned. If there is no path, the connection will not be made. The implementation of this business logic can be based on the shortest path analysis method, and the difficulty lies in how to orderly output the path between two devices. In R&D, the shortest path algorithm is used to query whether there is a path, and the path information is returned sequentially through a series of set operations.

5.5 Application Results

The electric grid data modeling method studied in this paper has been applied in many provincial power companies, mainly in electric power operation, power maintenance, power equipment management, equipment information communication, and other business areas. The construction of power data model is the key to realize "digital" power grid, and is also an important part of energy management system and distribution management system. Taking the calculation result of line transformer relationship of a developed provincial power grid as an example, topology analysis and calculation are carried out for the power grid network with nearly one million devices and three million devices connected. The research results in this paper can reduce the calculation time of line transformer relationship from 90 min to less than 30 min, and control the error rate of topology calculation within 3%. Therefore, it is of great theoretical value to study the construction technology of power system data model.

The power grid data model established in this paper can meet the requirements of complex and changeable power system with strong integrity, and can meet the requirements of accurate, convenient and fast power business analysis, which will greatly improve the level of power grid business departments in digital applications and business expansion.

Acknowledgements This work was supported by State Grid Corporation of China's Science and Technology Project (5400-202258431A-2-0-ZN) which is 'Research on deep data fusion and resource sharing technology of new distribution network'.

References

1. Feng, R.H., Zhao, Z., Xie, S., Huang, J.L., Wang, W.: Low voltage distribution network topology identification method based on principal component analysis and convex optimization. J. Tianjin Univ. **54**(7), 746–753 (2021). (in Chinese)
2. Li, Y.Z., Zhao, Q.S., Liang, D.K., et al.: Construction of low-voltage distribution network impedance topology model based on measurement information. Power Grid Clean Energy **37**(4), 15–22 (2021). (in Chinese)
3. Yan, Y.H., Zhou, R.X., Yu, W., et al.: Station topology identification based on carrier wave and power frequency distortion technology. Autom. Technol. Appl. **40**(10), 68–71 (2021). (in Chinese)
4. Yan, W.G., Wang, G.M., Lin, J.K., et al.: Verification method of low-voltage distribution network topology based on AMI measurement information. China Power **52**(2), 125–133 (2019). (in Chinese)
5. Sano, K., Takasaki, M.: A surgeless solid-state DC circuit breaker for voltage-source-converter-based HVDC systems. IEEE Trans. Ind. Appl. **50**(4), 2690–2699 (2014)
6. Magnusson, J., Bissal, A., Engdahl, G., Martinez-Velasco, J. A.: Design aspects of a medium voltage hybrid DC breaker. In: Innovative Smart Grid Technologies Conference Europe (ISGT-Europe), pp.1–6. IEEE, Istanbul (2014)
7. Zhang, B., Zhao, Y.H., Tang, L., et al.: Research on the verification method of urban low-voltage distribution network topology based on AMI data. Electron. Meas. Technol. **42**(6), 67–71 (2019). (in Chinese)
8. Zhang, L.Q.: Research on topology identification of power supply network based on smart meter data. Shandong University, Jinan (2020). (in Chinese)

9. Zhou, Y.Z., Wu, J.Y., Ji, L.Y., et al.: Two-stage support vector machines for transient stability prediction and preventive control of power systems. Proc. CSEE **38**(1), 137–147 (2018). (in Chinese)
10. Moulin, L.S., Da Silva, A.A., El-Sharkawi, M.A., Marks, R.J.: Support vector machines for transient stability analysis of large-scale power systems. IEEE Trans. Power Syst. **19**(2), 818–825 (2004)

Chapter 6
Improving CXR Self-Supervised Representation by Pretext Task and Cross-Domain Synthetic Data

Shouyu Chen, Yin Wang, Ke Sun, and Xiwen Sun

Abstract Supervised deep learning techniques have facilitated Chest X-ray (CXR) classification. Transfer learning from ImageNet pre-trained weights has become a common practice in medical image analysis. However, it may be suboptimal in the CXR setting because of distribution disparity, reducing the quality of the learned representations. On the other hand, mining deep features from the non-annotated image using self-supervised learning has great potential for medical tasks. This paper studies the influence of the self-supervised pretext task and non-annotated data amount for CXR self-supervised representation learning. We design a domain-specific self-supervised pretext task in the form of a data augmentation pipeline and explore the feasibility of using CXR and even computerized tomography (CT) volumes to expand the non-annotated CXR database. We verify our method on two state-of-the-art (SOTA) self-supervised architectures, BYOL and SimSiam, and report the results on two public datasets, Xray14 and COVID-QU-Ex. Our main findings are (1) XR-Augment, the proposed data augmentation, outperforms its counterparts in SOTA architectures on the CXR datasets; (2) based on cross-domain ImageNet pre-trained weights, self-supervised learning with XR-Augment further improves the discriminability of model weights on the CXR, and more non-annotated CXR enlarges this advantage; and (3) the synthesized pseudo-CXR, generated from CT, also helps in the context of CXR self-supervised learning.

S. Chen (✉) · Y. Wang
Department of Computer Science and Technology, College of Electronics and Information Engineering, Tongji University, Shanghai, China
e-mail: 1910667@tongji.edu.cn

K. Sun
Department of Radiology, Huashan Hospital, Fudan University, Shanghai, China

X. Sun
Department of Radiology, Shanghai Pulmonary Hospital, Tongji University School of Medicine, Shanghai, China

S. Patnaik et al. (eds.), *3D Imaging—Multidimensional Signal Processing and Deep Learning*, Smart Innovation, Systems and Technologies 348,
https://doi.org/10.1007/978-981-99-1145-5_6

6.1 Introduction

Deep learning facilitates the development of medical image classification techniques. CXR is an ordinary data modality in clinical diagnosis. The publicly available CXR dataset ChestX-ray14 (abbreviated as Xray14) [1] contains 14 thoracic diseases with labels automatically extracted by natural language processing (NLP). The COVID-QU-Ex dataset [2] is currently the largest COVID-19 CXRs public dataset. It contains annotations of three classes. However, the number of samples in the above datasets is still much less than that of ImageNet because of the costly labor power that requires expert knowledge. Therefore, using ImageNet pre-trained weights to initialize the neural network remains a practical choice.

The self-supervised learning (SSL) is one of the potential solutions which can further extend the advantages of transfer learning. One of the self-supervised methods is contrastive learning. It typically improves the discriminability of learned representations by specific-designed pretext tasks. One form of the pretext task is task-relevant data augmentation. Although some recent works proposed transformations on natural image contrastive learning, there is currently no consensus on medical ones [3]. Some discrepancies exist between CXR and natural images, the smaller size of primary visual features, regular morphological arrangement of anatomic structure, and the different number of image channels. The average share of an object (lesion) to the image in the Xray14 dataset, 7.54%, is much lower than in ImageNet. Because of several pooling layers in CNNs, only a few features from the small-sized region of interest can arrive at the output layer of the neural network. One viable solution for increasing the learned feature discriminability is to promote the transforms in data augmentation.

Besides, it is natural to combine supervised pre-training on a large-scale out-of-domain dataset, self-supervised pre-training on an in-domain dataset, with supervised training on the target dataset sequentially, also known as hierarchical pre-training (HPT) [4]. HPT is a promising solution to facilitate transferring features learned from large-scale natural image datasets to the data-hungry CXR domain. In this process, increasing the amount of non-annotated data used for self-supervised pre-training may be a wise choice to improve the quality of the final representation. A more ambitious practice is collecting non-annotated samples from out-of-domain synthesized data (i.e., generating CXR from CT), whose impact on contrastive learning is unknown.

We identify two challenges with the self-supervised CXR representation learning in the HPT paradigm: (1) Domain-inappropriate pretext task, lowering the efficiency of feature learning on CXR. (2) Inadequate non-annotated CXR for self-supervised learning. The main contributions of this paper are as follows.

1. We propose a CXR-specific data augmentation pipeline, XR-Augment, as a pretext task. It is suitable for different self-supervised contrastive learning architectures and CXR datasets.

2. We quantitatively evaluate the influence of the data amount of non-annotated CXRs (and pseudo-CXRs) on self-supervised feature discriminability by linear evaluation, thus confirming their values for medical image processing.
3. Our final model reached a macro-AUC score of 0.8036 on the Xray14 dataset and a macro-F1-score of 0.9116 on COVID-QU-Ex dataset, surpassing existing SOTA methods.

6.2 Related Works

6.2.1 Overview of CXR Classification

Now, two technology families, network architecture search (NAS) and convolutional neural networks (CNNs), dominate the Xray14 task. NAS usually uses multi-objective optimization algorithms. Reference [5] maintained the highest macro-AUC (0.847) by task-dependent NAS design. It reduced the computation cost and balanced model accuracy and deployment requirements. Reference [6] compacted the model while maintaining accuracy. CNNs family generally fine-tunes ImageNet pre-trained weights on the target dataset [7] because of lacking annotated samples. Reference [8] investigated 15 CNNs on two CXR datasets and demonstrated that small-sized custom models comparable to the larger ones. Reference [9] revealed that ImageNet pre-training can enhance the performance of all 16 common CNNs on the CXR interpretation task, and small-sized models benefit more. The above works indicate that ImageNet pre-training matters for CNNs. As few annotated data is available for supervised training, self-supervised learning advanced CXR tasks in recent years and is promising in improving the performance of CNNs on the CXR datasets.

6.2.2 Self-Supervision and Contrastive Learning

Self-supervised learning became popular because of its ability to mine features directly from non-annotated data [10]. Unsupervised learning can better balance effectiveness and efficiency when there is a lack of formulation of a fair and equitable definition, which is the case when there are no labels. Contrastive learning is a self-supervised technique that trains a model by forcing similar samples close to each other in feature space and dissimilar ones far away. To capture instance similarity, [11] learned discriminative features at the instance level, separate itself from the supervised methods that differentiate features at the class level. Reference [12] showed that for self-supervised models, its transfer learning generalization capability can be improved log-linearly to dataset size, and the representations quality increases in parallel with the model capacity and the problem complexity. In the field of computer vision, some classical contrastive learning methods nowadays include but are not limited to the MOCO family [13, 14], SimCLR family [15, 16], BYOL family [17, 18], and SimSiam [19].

There are several contrastive learning methods in medical image processing regarding architecture, pretext tasks, and cross-domain data processing. Utilizing the characteristic of high dimensionality in 3-dimensional (3d) magnetic resonance imaging (MRI), [20] extended the granularity of contrastive learning to discover global–local medical features. Grounded on the target task, [21] proposed five 3d pretext tasks and used 3d Contrastive Predictive Coding [22] to learn self-supervised representations from MRI and CT. To recognize COVID-19 from chest CT, [23] used contrastive learning to learn initial weights from non-annotated CT. However, they only concentrate on the same data modality. Integrating different datasets is a typical data collection method for self-supervised CXR pre-training [3] . Reference [24] further proposed an omni-supervised learning method to train the CXR-diagnostic model using all available supervision signals, inspiring us to learn a self-supervised network with different modalities. Reference [25] train COVID-19 classification and segmentation models with real CXR and CT-generated CXR demonstrated the potential of cross-modality data in the CXR task. Nevertheless, it only studied supervised settings. Last but not least, self-supervised fine-tuning of the ImageNet pre-trained model by the self-supervision stage in the HPT paradigm can improve the domain adaptability of the pre-trained weights on CXR.

6.2.3 Pretext Task and Data Augmentation

Pretext task is a crucial self-supervised task in contrastive learning. Its principal purpose is learning a model invariant to transformations applied to raw data points while remaining discriminative to other samples. Contrastive Predictive Coding [22] learned representations by predicting the future in latent space using autoregressive models. Reference [26] trained network to recognize the rotation applied to the image that the model gets as input. However, the bias introduced through self-supervised transformations could be a double-edged sword, as the learned transformation invariance could not be beneficial in all cases. For instance, [27] revealed that rotation does not work well for the image constructed by color textures, which is the case in the DTD dataset [28]. This shortcoming indicates the necessity of designing pretext task according to a specific data domain or modality.

Inspired by [14, 15, 26, 29, 30], data augmentation can function as a pretext task. Experiments in their works demonstrated that selecting robust pretext tasks or suitable data augmentations can highly boost the discriminability of the representations. Likewise, SwAV [31] outperformed other methods by numerous augmentations. Reference [21] highlighted the necessity to design the data augmentation for the specific task to improve the final model performance. To ensure proper data distribution, [32] checked the validity of data augmentation transforms according to the given dataset. Based on domain-specific knowledge, [33] proposed a pretext task of cross-stain prediction, introducing randomness in the Hematoxylin and Eosin Stain separation function to generate diverse data.

Designing fine-grained data augmentation as a pretext task is a natural solution for self-supervised learning and has not been well studied for CXR. Continuing this line of thought, a random combination of valuable transformations is feasible. AutoAugment [34] used the reinforcement learning technique to simplify hand-tuned data augmentation, but it is inconvenient to control the magnitude of the transformation. RandAugment [35], with lower complexity, outperformed the former by randomly sampled data augmentation transforms at each training iteration (step) under a preset magnitude. Notably, in unsupervised [15, 30] and semi-supervised [36] learning, neural networks benefited more from higher data augmentation magnitude than the supervised ones, demanding magnitude-controllable data augmentation. It again inspired us to re-design the data augmentation pipeline for CXR self-supervised tasks because the commonly used ones are natural image-specific.

6.3 Problem Definition

We first formulate our work as self-supervised contrastive learning, stage II of the HPT paradigm. Our target is to learn CXR self-supervised representations of better discriminability by the specific-designed pretext task and more non-annotated CXR. Finally, we evaluate the quality of the learned representation by the supervised (multi-label or single-label) multi-class linear evaluation on the target datasets.

6.3.1 Contrastive Learning Pretext Task

The pretext task in this paper is instance discrimination. An instance is the augmented view of a CXR image. Given the non-annotated training dataset of the contrastive learning $D_u = \{x^{(1)}, x^{(2)}, \dots, x^{(N_u)}\}$, where N_u is the number of CXRs. We choose a Siamese network, BYOL [17], as self-supervised contrastive learning architecture. It consists of two independent branches, student and teacher. Structurally, the student contains three components, backbone ϕ_S, projector φ_S, and predictor ψ_S. In contrast, the teacher has only a backbone ϕ_T and a projector φ_T. A_S and A_T are the data augmentation pipelines for the student and teacher branches, respectively. $\tau_S \sim A_S$ and $\tau_T \sim A_T$ are the data augmentation instances sampled in each training step. The model learns discriminative representations from the two augmented views, $\left(v^{(i,1)}, v^{(i,2)}\right) \triangleq \left(\tau_S\left(x^{(i)}\right), \tau_T\left(x^{(i)}\right)\right)$, of each CXR $x^{(i)} \in D_u$, that is, whether the views come from the same CXR. We denote the output of the student as $z_1^{(i,1)} \triangleq \psi_S\left(\varphi_S\left(\phi_S\left(v^{(i,1)}\right)\right)\right)$ and the output of the teacher as $z_2^{(i,2)} \triangleq \varphi_T\left(\phi_T\left(v^{(i,2)}\right)\right)$. The loss function is negative cosine similarity $\mathrm{S}\left(z_1^{(i,1)}, z_2^{(i,2)}\right) \triangleq -\frac{z_1^{(i,1)}}{\|z_1^{(i,1)}\|} \cdot \frac{z_2^{(i,2)}}{\|z_2^{(i,2)}\|}$, where $\|\cdot\|$ is the L2-norm. Following [20], we train the Siamese network using

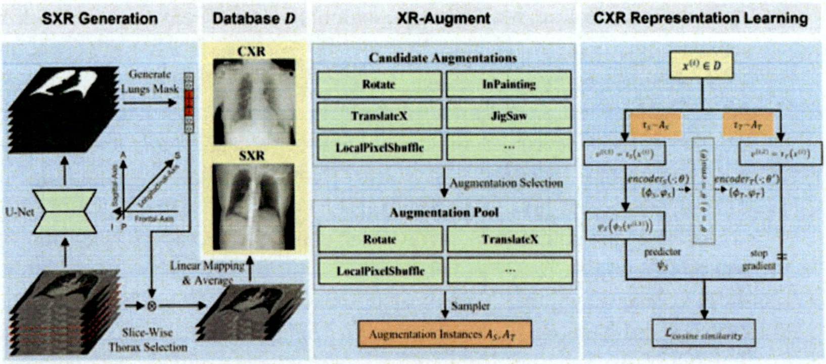

Fig. 6.1 Overview of the methods presented in this paper. **SXR Generation** explains the process of generating CT-synthesized CXRs (SXRs) from CT volumes to construct the non-annotated **Database** ***D***. **XR-Augment** demonstrates the construction of a CXR-oriented data augmentation pipeline consisting of suitable transforms to be used in the proposed self-supervised contrastive learning framework, **CXR Representation Learning**, shown in the right panel. The weight share in SimSiam is denoted as $\theta' = \theta$, and $\theta' = ema(\theta)$ is the exponential moving average update in BYOL. After training, the self-supervised pre-trained backbone ϕ_s can extract embeddings from the target dataset

a symmetric loss, formulated as $\mathcal{L} = \frac{1}{2}\left[S\left(z_1^{(i,1)}, z_2^{(i,2)}\right) + S\left(z_2^{(i,1)}, z_1^{(i,2)}\right)\right]$, where $z_2^{(i,1)} \triangleq \psi_S(\varphi_S(\phi_S(v^{(i,2)})))$, and $z_1^{(i,2)} \triangleq \varphi_T(\phi_T(v^{(i,1)}))$. Note that we update ϕ_T and φ_T by exponential moving average of ϕ_S and φ_S, but ϕ_S, φ_S, and ψ_S by gradient descent, same as [17]. We additionally use another SOTA model, SimSiam, in our experiments, and it updates the parameters of ϕ_T and φ_T by weight sharing with ϕ_S and φ_S. We show the workflow in the right panel of Fig. 6.1. After training, we discard all modules except ϕ_S.

6.3.2 *Supervised Multi-class Linear Evaluation*

In linear evaluation, $D_s = \{(x^{(1)}, y^{(1)}), \ldots, (x^{(N_s)}, y^{(N_s)})\}$ represents the training set of the supervised target task, where N_s is the number of samples. The supervised model $f = \{\phi, h\}$ consists of a backbone ϕ and a linear classification head h. We initialize ϕ by the trained backbone ϕ_S and h by random initialization. f receives a CXR-label pair, $\{x^{(i)}, y^{(i)}\} \in D_s$, and outputs corresponding prediction $\hat{y}^{(l)} = f(x^{(i)})$. The loss $\mathcal{L}\left(y^{(i)}, \hat{y}^{(l)}\right)$ is optimized only concerning h by a gradient descent optimizer.

6.4 Method

We improve the self-supervised CXR representation from two aspects, pretext task and data amount of non-annotated images. This section introduces (1) Selection of Candidate Transformations, which describes choosing data augmentation transforms suitable for CXR modality from candidate ones, (2) XR-Augment (Fig. 6.1, middle panel), a CXR-specific data augmentation pipeline that serves as a pretext task to generate augmented views, applicable to both CXR and synthetic X-ray (SXR), and (3) Pseudo-CXR Generation, a method to generate SXR from thoracic CT volume to expand the non-annotated CXR dataset, shown in Fig. 6.1, left panel.

6.4.1 Selection of Candidate Transformations

Following the principle of designing data augmentation according to the specific task [21], we first collect data augmentation candidate transformations from the related works of literature, of which the computation process should be meaningful for grayscale CXR. Then, we select the ones beneficial to the discriminability of the self-supervised features and append them to the end of the base data augmentation list (pipeline).

The commonly used four types of transforms are intensity-based, geometric-based, context-based, and cross-modal-based (view prediction). The first three of the above are relevant to our work. We group the collected transforms as follows.

1. Intensity-based: AutoContrast [35], Contrast [35], Brightness [35], Equalize [35], Posterize [35], Sharpness [35], Bézier curve [37], Gaussian noise [15], and Sobel filtering [15].
2. Geometric-based: Rotate [35], ShearX [35], ShearY [35], TranslateX [35], and TranslateY [35].
3. Context-based: LocalPixelShuffle [37], InPainting [37], OutPainting [37], and JigSaw [38].

To facilitate the effective feature transfer within the neural network, we expand the perceptual field of the base data augmentation pipeline by setting the hyperparameter scale in RandomResizedCrop() to (0.72, 1), whose default assignment is (0.08, 1). Following is the selection process for each candidate transform.

1. Let $L \triangleq \big[\text{RandomResizedCrop(scale} = (0.72, 1)), \text{RandomHorizontalFlip}\big]$, $L_{\text{base}} \triangleq L \cup F_I$ the base data augmentation pipeline for both branches in contrastive learning architecture, where F_I indicates identity transformation. We determine the hyperparameter **scale** by ablation experiments. The detailed result is in Fig. 6.2. We denote the linear evaluation score of backbone ϕ, self-supervised pre-trained with L_{base} on the Xray14, as S_{base}.
2. We regenerate L by substituting F_I in L_{base} with each candidate transformation, F_{tr}, by turns, re-training contrastive learning model with L, and then comparing

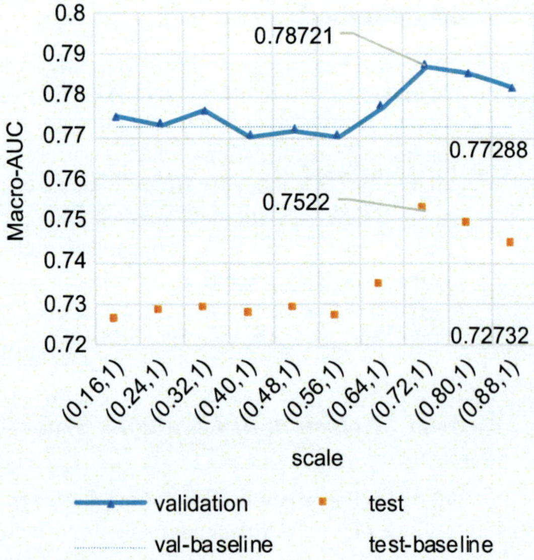

Fig. 6.2 Influence of input view size of data augmentation on the self-supervised representations. On top of the base data augmentation pipeline L_{base}, we merely tune the scale in the RandomResizedCrop(). scale = (0.72, 1) is the optimum point in the Xray14 setting

the derived linear evaluation score S of ϕ with the S_{base}. We use an ablation study to find the best assignment of its hyperparameter for each transformation. Please refer to Fig. 6.3 for complete results. After that, we will get a score sequence $\{S_i\}$ under different assignments $\{h_i\}$, $h_i \in H$, and H is the set of possible hyperparameter assignments for F_{tr}. We save F_{tr} to the transformation pool L_{pool} if its $\max(\{S_i\}) > S_{\text{base}}$.

Collectively, we pinpoint the objective of the selection process for each candidate F_{tr} as

$$argmax_{h_i \in H} \Psi(\phi(F_{\text{tr}}(h_i), D_u), D_s), s.t. \exists h_i \in H, \Psi(\phi(h_i)) > \Psi(\phi(h_{\text{base}})) \quad (6.1)$$

For a given data augmentation, F_{tr}, $\phi(F_{\text{tr}}(h_i), D_u)$ indicates the backbone, trained on D_u, with data augmentation transformation F_{tr} parameterized by h_i, and $\Psi(\cdot, D_s)$ indicates the linear evaluation on dataset D_s. Proof: h_{base} represents the hyperparameter assignment that will be compared against, excluding using F_{tr}.

6.4.2 *XR-Augment*

To adapt the existing contrastive learning architecture to CXR, we re-design a CXR-specific data augmentation pipeline with the selected domain-specific transformations above that helps in feature discriminability. As [35], we treat identity mapping as an independent transformation, i.e., $L_{\text{pool}} \triangleq L_{\text{pool}} \cup F_I$. We design a weighted

Fig. 6.3 Ablation results for all candidate transformations. The baseline scores are from the scale = (0.72, 1) in Fig. 6.2. For each transform, if its validation or test score exceeds the benchmark, we mark its highest score using a numerical label

sampling strategy for each transform in L_{pool} by considering its relative importance concerning the benchmark score, calculated by the best growth of each transformation against the S_{base}. Given the performance score vector $\boldsymbol{s}$, each item within it represents the best score of each $F_{tr} \in L_{pool}$. Below is the calculation of normalized sampling weights.

$$w = \text{softmax}(s - s_{\text{base}}) \tag{6.2}$$

Then, we build the data augmentation pipeline $XR - Augment\left(L_{\text{base}}, L_{\text{pool}}, \boldsymbol{w}, N\right)$. N is the number of transforms sampled in each training step. We use a symmetric data augmentation design in the self-supervised model. Namely, two branches of the Siamese-based model employ the same data augmentation pipeline, same as SimSiam [19] but different from BYOL [17]. For each $F_{tr} \in L_{\text{pool}}$, its hyperparameter is preset according to its best performance in the ablation study. We detail XR-Augment for one data batch in Algorithm 1.

Algorithm 1 XR-Augment

Input: A batch of CXR images, $\mathcal{X} = \{x^{(i)}\}$ $(i = 1, ..., N_{bs})$, with size N_{bs}
Parameter: Base transformation list L_{base}, transformation pool L_{pool}, weights vector $\boldsymbol{w}$, number of transformations to sample in each training step N
Output: Augmented CXR view pairs $\mathcal{X}' = \{(v^{(i,1)}, v^{(i,2)})\}$ $(i = 1, ..., N_{bs})$

$\mathcal{X}' \leftarrow EmptyList$;
for each $x \in \mathcal{X}$ **do**
    $L_{tr} \leftarrow Sample(L_{pool}, \boldsymbol{w}, N, replace = True)$; ▷ Downsampling.
    $L \leftarrow L_{base} \cup L_{tr}$; ▷ Function set for the current iteration.
    $v^1 \leftarrow x$;
    $v^2 \leftarrow x$;
    for each $F_{tr} \in L$ **do** ▷ Iterate data augmentation operators.
        $f \sim F_{tr}$; ▷ Generate an instance of data augmentation.
        $v^1 \leftarrow f(v^1)$;
        $g \sim F_{tr}$;
        $v^2 \leftarrow g(v^2)$;
    end for
    $\mathcal{X}' \leftarrow \mathcal{X}' \cup (v^1, v^2)$; ▷ Append tuple of augmented view pair to list.
end for
return $\mathcal{X}'$

6.4.3 Pseudo-CXR Generation

This subsection introduces the generation procedure of synthetic pseudo-CXR (abbreviated as SXR) in the anterior–posterior (AP) view from CT data to expand the non-annotated CXR database. Given a patient's CT volume (3d tensor), we generate SXR as the following process.

1. **Mask and re-sample.** Extract the lungs' mask by the pre-trained U-Net [39] and re-sample the original tensor to the uniform spatial resolution ($1mm$) in all three dimensions.
2. **Thorax Selection.** Remove all slices along the AP-axis (Sagittal View) that do not cover the lungs' region. We keep an additional 30 slices of the outer lung along the AP-axis for complete coverage of the thoracic region.
3. **Rotation.** In order to enrich the representation diversity of volume. We rotate around the Longitudinal (Superior-Inferior) Axis that crosses the center of the lungs according to the angles with stride specified by range$(-15, 16, 5)$. The Sagittal Axis and Frontal Axis by range$(-10, 11, 5)$. We then get a permutation of angles at a time by traversing the 3d angle grid. We use python pseudo-code here to represent the angle series with stride.
4. **Normalization.** The tensor is clipped to the interval $[-1100, 2500]$ using hyper-parameters WINDOW LOCATION = 700 and WINDOW WIDTH = 1800, which makes the final SXR visually similar to a CXR. It is then linearly mapped into the interval $[0, 255]$.
5. **Imaging.** We average the final tensor along the Sagittal Axis to reformat it into a 2-dimensional matrix and save it as an SXR file.

All the generated data constitute the SXR data pool. We exhibit three SXR samples in Fig. 6.4, line 2, and the overall workflow in the left panel of Fig. 6.1.

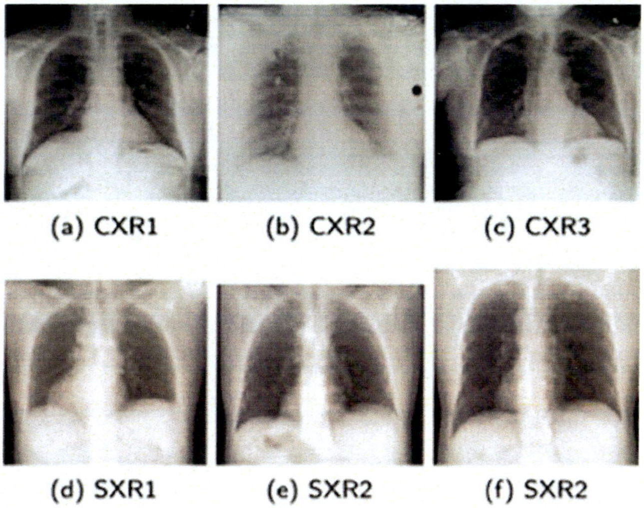

Fig. 6.4 Examples of CXRs and SXRs. The first row exhibits three CXR samples from the Xray14 dataset, and the second row shows the SXRs generated from our in-house, DSB, and LIDC-IDRI datasets, respectively

6.5 Experiment

6.5.1 *Data*

Target dataset. We use two public thoracic X-ray classification datasets, Xray14 [1] and COVID-QU-Ex [2]. Xray14 is a multi-label multi-class radiographic thoracic disease classification dataset. It contains 112,120 CXRs, each with 15 labels indicating the existence of 14 symptoms and No Findings. Each CXR has 1.26 annotations on average. Its official training set and test set contain 86,524 and 25,596 CXRs in several. We randomly split the official training set into a training set containing 75,774 (denoted as $N_{\mathbf{data}}$) CXRs and a validation set containing 10,750 CXRs in the ratio of 7:1 at the patient level, with same as [1, 7]. We detail the split in line 2, Table 6.1. During the linear evaluation, we use only annotations of 14 symptom classes to train the model (head module h) because the No Finding is redundant knowledge about the existence of 14 diseases.

COVID-QU-Ex is a single-label multi-class classification dataset containing three classes, Normal, Non-COVID, and COVID-19. It has 33,920 CXRs, including 11,956 COVID-19, 11,263 Non-COVID infections (Viral or Bacterial Pneumonia), and 10,701 Normal CXRs. We use its official dataset splits, in which the training set contains 21,716 CXRs, validation set 5,417, and test set 6,533.

Contrastive learning dataset. We use our Xray14 training set split **without annotations** as the base dataset. CXRs with AP or PA views from CheXpert [40] and PadChest [41] serve as in-domain non-annotated external data and are thereupon used to construct non-annotated datasets with the size of $2N_{data}$ and $4N_{data}$ separately (refer to the last line in Table 6.1 for the details).

CT dataset. For the SXR generation, we use two public CT datasets, LIDC-IDRI [42] (1018 volumes) and DSB17 [43] (2101 volumes), and one in-house CT dataset (506 volumes) from the Pulmonary Hospital of Tongji University. We generated 634,375 (= 3625 × 7 × 5 × 5) SXR images, namely the SXR data pool. Similar to CXR in Table 6.1, we sample SXRs with a size of N_{data} to $4N_{\mathrm{data}}$ from the SXR data pool for the SXR experiment (Fig. 6.6).

Table 6.1 Dataset splits of Xray14 supervised (Sup.) and CXRs self-supervised (Self-Sup.) experiments. AP denotes anterior–posterior and PA denotes posterior-anterior

Data source	Xray14 official test set #25,596	Xray14 official train set #86,524		PadChest (AP&PA views) # 63,848		CheXpert (AP&PA views) # 191,229
Sup.	Test set #25,596	Validation set #10,750	Train set #75,774	N/A		
Self-Sup.	N/A		(N_{data}) #75,774	(N_{data}) #75,774	(N_{data}) #75,774	Not use #27,755

6.5.2 Settings

We choose ResNet-50 (abbreviated as R50) [44] as the backbone and normalize all CXRs and SXRs by the statistics of Xray14, mean = 0.506, and std = 0.252, then resize them to 224 × 224 resolution as model input. Before using ImageNet [45] pre-trained weights, R50's first convolution layer weights are averaged across the channel dimension to adapt the 1-channel inputs. We implement the distributed data-parallel training by the PyTorch and MMSelfSup [46] library. Unless otherwise stated, the settings of supervised and self-supervised experiments are as follows.

Self-supervised pre-training experiments. We chose BYOL [17] as the default contrastive learning framework. For some experiments, we also report the results of another SOTA contrastive learning model SimSiam [19]. They are both SimSiam-based networks with negative cosine similarity [17] loss function. We use Nesterov-LARS [47] optimizer with the following configurations: batch size = 512, a linearly scaled learning rate $\text{lr} = 3\text{e}^{-1} \times 512 \div 256$ for BYOL and $\text{lr} = 1\text{e}^{-1} \times 512 \div 256$ for SimSiam, weight decay = 1e^{-6}, and Nesterov momentum = 0.9. In our SimSiam hierarchical pre-training experiments, the learning rate of the encoder part is scaled by $\times 0.1$ to ensure convergence. We train each self-supervised model for 40 epochs, in which the first two epochs employ linear learning rate warm-up, and the rest epochs schedule the learning rate by cosine annealing. The rest settings remain consistent with [17].

Supervised linear evaluation. We use Nesterov-SGD optimizer with learning rate $\text{lr} = 3\text{e}^{-1} \times 512 \div 256$, momentum = 0.9, weight decay = $1e^{-6}$, and choose the multi-label asymmetric loss ASL [48] as the loss function for Xray14, with $\gamma^{+} = 0$, $\gamma^{-} = 2$ and clip = 0.01, and cross-entropy loss for COVID-QU-Ex. We trained the model for 15 epochs with batch size = 512. To ensure convergence, we scale lr by $\times 0.1$ at epochs 6 and 12. Finally, we select the best model which achieves the highest performance on the validation set.

Performance Metrics. 1) Metrics on self-supervised representations. The objective of self-supervised learning is to learn a model invariant to transformations on the input data while maintaining inter-sample discriminability. We measure the 1st-Wasserstein distance on the self-supervised features to evaluate the transformation invariance. The backbone features are generated from the augmented view pairs of each sample by the self-supervised pre-trained R50 backbone followed by global average pooling. We get the averaged distance over all data points. 2) Metrics on linear evaluation. Since the class distributions of the official training and the test set of Xray14 differ significantly, to focus on comparing the self-supervised feature discriminability, we are mainly concerned with macro-AUC scores on our same-distributed validation set, treat it as the primary evaluation metrics, same with [14]. For a comprehensive comparison, we also report the test set scores. For the COVID-QU-Ex dataset, we select the macro F1-score on the official validation set and test set as metrics.

Table 6.2 Linear evaluation with different initialization weights on Xray14 validation and test sets. We use two SOTA self-supervised contrastive learning architectures, BYOL and SimSiam, and their default implementation, with the only exception of reducing the number of input channels of the first convolutional layer. We denote random initialization as RI, ImageNet out-of-domain supervised pre-training as IN, Xray14 self-supervised contrastive learning pre-training as SSL, and IN followed by SSL as HPT

Init.	Val/Test macro-AUC	
	BYOL	SimSiam
RI	52.438/51.744	
IN	74.721/71.281	
SSL	54.089/52.26	72.549/68.235
HPT	**77.288**/72.732	**76.877**/73.483

6.5.3 *Result and Analysis*

HPT Improving CXRs Initial Representations. We first verify the feasibility of hierarchical pre-training on Xray14 by comparing different weights initialization methods in Table 6.2. The control subjects include random initialization, ImageNet supervised pre-training, Xray14 self-supervised pre-training, and ImageNet supervised pre-training followed by Xray14 self-supervised pre-training. This experiment uses CXRs with the size of N_{data} in SSL training. We can see that score improves upon random initialization by either ImageNet supervised pre-training or CXRs self-supervised pre-training. Notably, we achieve the highest performance using both supervised and self-supervised pre-training sequentially, namely hierarchical pre-training. This observation is consistent with two SOTA self-supervised models, BYOL and SimSiam, and validates the importance of the hierarchical pre-training paradigm on Xray14.

XR-Augment. By ablation study, we first determine the optimal view size (hyperparameter scale) for the RandomResizedCrop() function in the Xray14 self-supervised learning task. The experimental result in Fig. 6.2 shows that scale = (0.72, 1) is the optimum point on Xray14, and we use it as the default hyperparameter in the following experiments. Then, we use an ablation study for each candidate data augmentation transformation to determine whether to add it to the L_{pool}. The transforms preserved experimentally include Rotate, ShearX, ShearY, TranslateY, LocalPixelShuffle, InPainting, OutPainting, JigSaw, AutoContrast, Contrast, Brightness, Equalize, and Bézier curve. For more details, please refer to Fig. 6.3. We find the best assignment $N = 3$ for XR-Augment and show the result in Fig. 6.5.

We replace the default data augmentation pipeline with XR-Augment in BYOL and SimSiam in several. Table 6.3 shows that our method can consistently outperform the SOTA counterparts and verifies that **our domain-oriented XR-Augment can generalize to different contrastive learning models**, despite being designed on the BYOL. To better understand this result, we measure the 1st-Wasserstein distance on the features generated on two augmented views of one sample by the pre-trained

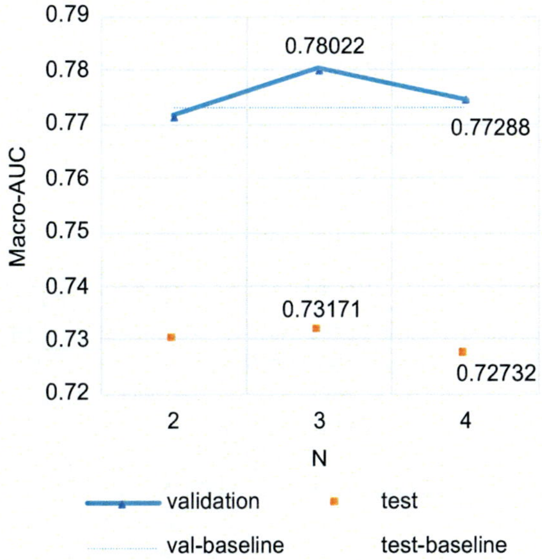

Fig. 6.5 XR-Augment ablation experiment. N is the number of randomly selected transformations. Benchmark scores are from the official data augmentation of BYOL

Table 6.3 The performance improvement of the proposed XR-Augment compared to the benchmark method on different SOTA self-supervised contrastive learning architectures

Aug.	Val/Test macro-AUC	
	BYOL	SimSiam
Official	77.288/72.732	76.877/73.483
Augment ($N = 3$)	**78.022**/73.171	**77.08**/73.622

Table 6.4 The 1st-Wasserstein distances measure the averaged similarity between two views' representations of each sample. The results come from BYOL pre-trained R50 backbone. Smaller values indicate better transformation invariance of the backbone

Aug. category	Distance
BYOL	0.0376
+XR-augment	0.0234 (−0.0142)
+4N_{data} (CXR)	0.0132 (−0.0102)
+4N_{data} (SXR)	0.0223 (−0.0011)

backbone. We list the results in rows 2 and 3 in Table 6.4. The solution that performs better in the linear evaluation of the target dataset possesses a shorter distance.

We also investigate the contribution of different data augmentation types in L_{pool} to the data representations, present the results in Table 6.5, and observe that the intensity-based transforms outperform the other two, which indicates that the self-supervised

Table 6.5 The relative contribution of different transformation types. We show the performance gains of each data augmentation type on the Xray14 and present the result as mean and standard deviation. For convenient observation, we scale the values by ×100

Aug. type	Mean ± Std.
Intensity-based	**0.5473 ± 0.2472**
Geometric-based	0.3062 ± 0.1464
Context-based	0.1743 ± 0.0648

contrastive learning stage of the hierarchical pre-training paradigm benefits most from the intensity-based transformation when the target task is Xray14.

Data Amount Experiment. First, we compare the discriminability of the Xray14 embeddings in terms of CXR data amount, N_{data}, $2N_{\text{data}}$, and $4N_{\text{data}}$ and show the result in the blue markers in Fig. 6.6. The score rises in parallel with the number of non-annotated CXRs. We then conduct experiments with random sampling (without replacement) from the SXR data pool and show results with the yellow markers. Again, it exhibits a trend of monotonically increasing. This result verifies that using only SXR in self-supervised pre-training can improve the feature discriminability. However, the SXR score only exceeds the self-supervised pre-training benchmark with the size of $2N_{data}$ and $4N_{data}$. We speculate that this phenomenon relates to the amount of CT volumes used because the number of CT cases is currently few (∼3000 patients) while the feature diversity introduced by rotating CT volumes is limited. Thus, we may improve the result with more CT data in future research.

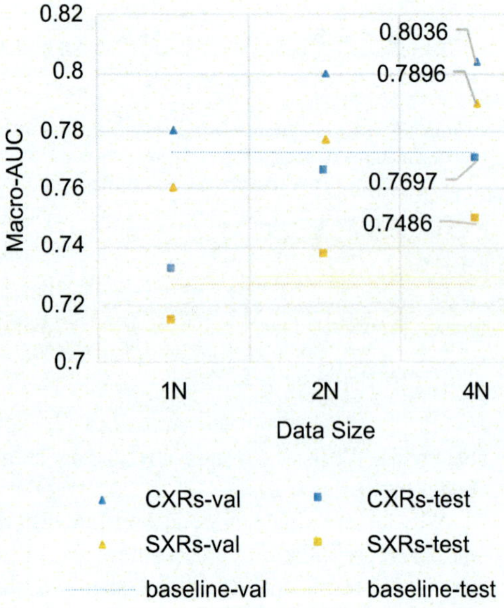

Fig. 6.6 The impact of non-annotated X-ray data amount on the self-supervised representations. The blue markers indicate the CXR results, and the orange markers indicate the SXR ones

Table 6.6 Summary of the linear evaluation on the COVID-QU-EX dataset

Aug. category	Distance
RI	0.5375/0.5545
+IN	0.7841/0.8040
+SSL (HPT)	0.8811/0.9103
+XR-augment	**0.8813**/0.8846
+4N_{data} (CXR)	**0.9116**/0.9279
+4N_{data} (SXR)	0.8578/0.8721

Results on COVID-QU-Ex. We also validate our method on the COVID-QU-Ex dataset and summarize the results in Table 6.6. From lines 3 to 6, performance improves consistently on the validation set. However, when discriminating COVID features, the SXR external dataset is not as effective as the SXR ones. We speculate this is related to the lower complexity of COVID-QU-Ex (fewer classes), higher baseline score, and the difference in feature distribution required to diagnose COVID-19 versus that of in Xray14, which makes it unable to transfer the self-supervision gains from SXR's data volume to the COVID-19 dataset. Even so, the $4N_{\text{data}}$ SXR results still outperform the fully supervised ImageNet pre-training benchmark (0.8721 vs. 0.8040), demonstrating the potential of SXR on the COVID-19 distribution.

6.6 Conclusion and Future Research

In this paper, we presented a synthesis of CXR data to increase the available data volume for CXR self-supervised pre-training tasks. We proposed XR-Augment, a CXR-specific data augmentation method, as a pretext task for contrastive learning with a downstream CXR classification task. Our quantitative analysis demonstrated its effectiveness in the self-supervised learning stage of hierarchical pre-training. Furthermore, we showed that using more data for pre-training can improve the representations learned by self-supervised models on CXRs and that synthesized SXR data can also be used as supplementary data for medical tasks that are data-hungry and have high annotation costs. This finding has the potential to inspire computer-aided clinical diagnosis. For future work, we plan to explore data augmentation in greater depth and design simplified and easy-to-use methods as pretext tasks for self-supervised learning. We also intend to refine our SXR generation procedure in a broader range of CT datasets.

References

1. Wang, X.S., Peng, Y.F., Lu, L., Lu, Z.Y., Bagheri, M., Summers, R.M.: Chestx-ray8: hospital-scale chest x-ray database and benchmarks on weakly-supervised classification and localization of common thorax diseases. In: Proceedings of the IEEE Conference on Computer Vision and Pattern Recognition, pp. 2097–2106 (2017). https://nihcc.app.box.com/v/ChestXray-NIHCC. Accessed 2 Dec 2022
2. Tahir, A.M., Chowdhury, M.E.H., Khandakar, A., Rahman, T., Qiblawey, Y., Khurshid, U., Kiranyaz, S., Ibtehaz, N., Rahman, M.S., Al-Maadeed, S., et al.: Covid-19 infection localization and severity grading from chest x-ray images. Comput. Biol. Med. **139**, 105002 (2021). https://www.kaggle.com/datasets/cf77495622971312010dd5934ee91f07ccbcfdea8e2f7778977ea8485c1914df. Accessed 15 Jan 2022
3. Dufumier, B., Gori, P., Victor, J., Grigis, A., Wessa, M., Brambilla, P., Favre, P., Polosan, M., Mcdonald, C., Piguet, C.M., et al.: Contrastive learning with continuous proxy meta-data for 3d MRI classification. In: International Conference on Medical Image Computing and Computer-Assisted Intervention, pp. 58–68. Springer (2021)
4. Reed, C.J., Yue, X.Y., Nrusimha, A., Ebrahimi, S., Vijaykumar, V., Mao, R., Li, B., Zhang, S.H., Guillory, D., Metzger, S., et al.: Self-supervised pretraining improves self-supervised pretraining. In: Proceedings of the IEEE/CVF Winter Conference on Applications of Computer Vision, pp. 2584–2594 (2022)
5. Lu, Z.C., Whalen, I., Dhebar, Y., Deb, K., Goodman, E.D., Banzhaf, W., Boddeti, V.N.: Multi-objective evolutionary design of deep convolutional neural networks for image classification. IEEE Trans. Evol. Comput. **25**(2), 277–291 (2020)
6. Lu, Z.C., Deb, K., Boddeti, V.N.: Muxconv: information multiplexing in convolutional neural networks. In: Proceedings of the IEEE/CVF Conference on Computer Vision and Pattern Recognition, pp. 12044–12053 (2020)
7. Rajpurkar, P., Irvin, J., Zhu, K., Yang, B., Mehta, H., Duan, T., Ding, D., Bagul, A., Langlotz, C., Shpanskaya, K., et al.: Chexnet: radiologist-level pneumonia detection on chest x-rays with deep learning (2017). arXiv:1711.05225
8. Bressem, K.K., Adams, L.C., Erxleben, C., Hamm, B., Niehues, S.M., Vahldiek, J.L.: Comparing different deep learning architectures for classification of chest radiographs. Sci. Rep. **10**(1), 1–16 (2020)
9. Ke, A., Ellsworth, W., Banerjee, O., Ng, A.Y., Rajpurkar, P.: Chextransfer: performance and parameter efficiency of imagenet models for chest x-ray interpretation. In: Proceedings of the Conference on Health, Inference, and Learning, pp. 116–124 (2021)
10. Jaiswal, A., Babu, A.R., Zadeh, M.Z., Banerjee, D., Makedon, F.: A survey on contrastive self-supervised learning. Technologies **9**(1), 2 (2020)
11. Wu, Z.R., Xiong, Y.J., Yu S.X., Lin, D.H.: Unsupervised feature learning via non-parametric instance discrimination. In: Proceedings of the IEEE Conference on Computer Vision and Pattern Recognition, pp. 3733–3742 (2018)
12. Goyal, P., Mahajan, D., Gupta, A., Misra, I.: Scaling and benchmarking self-supervised visual representation learning. In: Proceedings of the IEEE/CVF International Conference on Computer Vision, pp. 6391–6400 (2019)
13. He, K.M., Fan, H.Q., Wu, Y.X., Xie, S.N., Girshick, R.: Momentum contrast for unsupervised visual representation learning. In: Proceedings of the IEEE/CVF Conference on Computer Vision and Pattern Recognition, pp. 9729–9738 (2020)
14. Chen, X.L, Fan, H.Q., Girshick, R., He, K.M.: Improved baselines with momentum contrastive learning (2020). arXiv:2003.04297
15. Chen, T., Kornblith, S., Norouzi, M., Hinton, G.: A simple framework for contrastive learning of visual representations. In: International Conference on Machine Learning, pp. 1597–1607. PMLR (2020)
16. Chen, T., Kornblith, S., Swersky, K., Norouzi, M., Hinton, G.E.: Big self-supervised models are strong semi-supervised learners. Adv. Neural. Inf. Process. Syst. **33**, 22243–22255 (2020)

17. Grill, J.B., Strub, F., Altché, F., Tallec, C., Richemond, P.H., Buchatskaya, E., Doersch, C., Pires, B.A., Guo, Z.D., Azar, M.G., et al.: Bootstrap your own latent: a new approach to self-supervised learning (2020). arXiv:2006.07733
18. Richemond, P.H., Grill, J.B., Altché, F., Tallec, C., Strub, F., Brock, A., Smith, S., De, S., Pascanu, R., Piot, B., et al.: Byol works even without batch statistics (2020). arXiv:2010.10241
19. Chen, X.L., He, K.M.: Exploring simple siamese representation learning. In: Proceedings of the IEEE/CVF Conference on Computer Vision and Pattern Recognition, pp. 15750–15758 (2021)
20. Chaitanya, K., Erdil, E., Karani, N., Konukoglu, E.: Contrastive learning of global and local features for medical image segmentation with limited annotations (2020). arXiv:2006.10511
21. Taleb, A., Loetzsch, W., Danz, N., Severin, J., Gaertner, T., Bergner, B., Lippert, C.: 3d self-supervised methods for medical imaging (2020). arXiv:2006.03829
22. Oord, A.V.D., Li, Y., Vinyals, O.: Representation learning with contrastive predictive coding (2018). arXiv:1807.03748
23. Chen, X.C., Yao, L.N., Zhou, T., Dong, J.M., Zhang, Y.: Momentum contrastive learning for few-shot covid-19 diagnosis from chest CT images. Pattern Recogn. **113**, 107826 (2021)
24. Luo, L.Y., Chen, H., Zhou, Y.N., Lin, H.J., Heng, P.A.: Oxnet: Deep omni-supervised thoracic disease detection from chest x-rays. In: International Conference on Medical Image Computing and Computer-Assisted Intervention, pp.537–548. Springer (2021)
25. Tan, T., Das, B., Soni, R., Fejes, M., Ranjan, S., Szabo, D.A., Melapudi, V., Shriram, K., Agrawal, U., Rusko, L., et al.: Pristine annotations-based multi-modal trained artificial intelligence solution to triage chest x-ray for covid-19. In: International Conference on Medical Image Computing and Computer-Assisted Intervention, pp. 325–334. Springer (2021)
26. Komodakis, N., Gidaris, S.: Unsupervised representation learning by predicting image rotations. In: International Conference on Learning Representations (ICLR) (2018)
27. Yamaguchi, S., Kanai, S., Shioda, T., Takeda, S.: Multiple pretext-task for self-supervised learning via mixing multiple image transformations. CoRR (2019)
28. Cimpoi, M., Maji, S., Kokkinos, I., Mohamed, S., Vedaldi, A.: Describing textures in the wild. In: Proceedings of the IEEE Conference on Computer Vision and Pattern Recognition, pp. 3606–3613 (2014)
29. Chen, T., Zhai, X., Ritter, M., Lucic, M., Houlsby, N.: Self-supervised GANs via auxiliary rotation loss. In: Proceedings of the IEEE/CVF Conference on Computer Vision and Pattern Recognition, pp. 12154–12163 (2019)
30. Tian, Y., Sun, C., Poole, B., Krishnan, D., Schmid, C., Isola, P.: What makes for good views for contrastive learning? (2020). arXiv:2005.10243
31. Caron, M., Misra, I., Mairal, J., Goyal, P., Bojanowski, P., Joulin, A.: Unsupervised learning of visual features by contrasting cluster assignments. Adv. Neural. Inf. Process. Syst. **33**, 9912–9924 (2020)
32. Shorten, C., Khoshgoftaar, T.M.: A survey on image data augmentation for deep learning. J. Big Data **6**, 1–48 (2019)
33. Yang, P., Hong, Z., Yin, X., Zhu, C., Jiang, R.: Self-supervised visual representation learning for histopathological images. In: International Conference on Medical Image Computing and Computer-Assisted Intervention, pp. 47–57. Springer (2021)
34. Cubuk, E.D., Zoph, B., Mane, D., Vasudevan, V., Le, Q.V.: Autoaugment: learning augmentation strategies from data. In: Proceedings of the IEEE/CVF Conference on Computer Vision and Pattern Recognition, pp. 113–123 (2019)
35. Cubuk, E.D., Zoph, B., Shlens, J., Le, Q.V.: Randaugment: practical automated data augmentation with a reduced search space. In: Proceedings of the IEEE/CVF Conference on Computer Vision and Pattern Recognition Workshops, pp. 702–703 (2020)
36. Xie, Q., Dai, Z., Hovy, E., Luong, M.T., Le, Q.V.: Unsupervised data augmentation for consistency training (2019). arXiv:1904.12848
37. Zhou, Z., Sodha, V., Siddiquee, M.M.R., Feng, R., Tajbakhsh, N., Gotway, M.B., Liang, J.: Models genesis: Generic autodidactic models for 3d medical image analysis. In: International Conference on Medical Image Computing and Computer-Assisted Intervention, pp. 384–393. Springer (2019)

38. Noroozi, M., Favaro, P.: Unsupervised learning of visual representations by solving jigsaw puzzles. In: European Conference on Computer Vision, pp. 69–84. Springer (2016)
39. Hofmanninger, J., Prayer, F., Pan, J., Röhrich, S., Prosch, H., Langs, G.: Automatic lung segmentation in routine imaging is primarily a data diversity problem, not a methodology problem. Eur. Radiol. Exp. **4**(1), 1–13 (2020)
40. Irvin, J., Rajpurkar, P., Ko, M., Yu, Y., Ciurea-Ilcus, S., Chute, C., Marklund, H., Haghgoo, B., Ball, R., Shpanskaya, K., et al.: Chexpert: a large chest radiograph dataset with uncertainty labels and expert comparison. In: Proceedings of the AAAI Conference on Artificial Intelligence, pp. 590–597 (2019). https://stanfordmlgroup.github.io/competitions/chexpert/. Accessed 15 Jan 2022
41. Bustos, A., Pertusa, A., Salinas, J.M., de la Iglesia-Vayá, M.: Padchest: a large chest x-ray image dataset with multi-label annotated reports. Med. Image Anal. **66**, 101797 (2020). https://bimcv.cipf.es/bimcv-projects/padchest/. Accessed 15 Jan 2022
42. Armato, S.G., III., McLennan, G., Bidaut, L., McNitt-Gray, M.F., Meyer, C.R., Reeves, A.P., Zhao, B., Aberle, D.R., Henschke, C.I., Hoffman, E.A., et al.: The lung image database consortium (lidc) and image database resource initiative (idri): a completed reference database of lung nodules on CT scans. Med. Phys. **38**, 915–931 (2011)
43. Kaggle, B.: Kaggle data science bowl (2017)
44. He, K., Zhang, X., Ren, S., Sun, J.: Deep residual learning for image recognition. In: Proceedings of the IEEE Conference on Computer Vision and Pattern Recognition, pp. 770–778 (2016)
45. Deng, J., Dong, W., Socher, R., Li, L. J., Li, K., Fei-Fei, L.: Imagenet: a large-scale hierarchical image database. In: 2009 IEEE Conference on Computer Vision and Pattern Recognition, pp. 248–255. IEEE (2009)
46. Contributors, M.: MMSelfSup: OpenMMLab self-supervised learning toolbox and benchmark (2021). https://github.com/open-mmlab/mmselfsup. Accessed 15 Jan 2022
47. Ginsburg, B., Gitman, I., You, Y.: Large batch training of convolutional networks with layer-wise adaptive rate scaling (2018)
48. Ridnik, T., Ben-Baruch, E., Zamir, N., Noy, A., Friedman, I., Protter, M., Zelnik-Manor, L.: Asymmetric loss for multi-label classification. In: Proceedings of the IEEE/CVF International Conference on Computer Vision, pp. 82–91 (2021)

Chapter 7
Research on Dynamic Analysis Technology of Quantitative Control Oriented to Characteristics of Power Grid Digital Application Scenarios

Gang Wang, Jianhong Pan, Changhui Lv, Bo Zhao, and Aidi Dong

Abstract The digitalization construction of power grid companies will gradually be carried out in depth, while the research on the quantitative control dynamic analysis for the characteristics of power grid digitalization application scenarios has not yet been carried out. In this paper, the dynamic analysis technology of quantitative control of power grid is studied, and the expected project costs and benefits are preliminarily calculated using the functional analysis method; Based on expert scoring method, the quantitative influencing factor set of digital application scene target features is constructed; Based on fuzzy analytic hierarchy process, the quantitative impact index set of digital application scenario features is constructed; Based on K2 structure algorithm and Markov Monte Carlo structure algorithm, the application model of Bayesian network computing is constructed, and a dynamic technical and economic evaluation model adapted to different application scenarios is formed; Based on the calculation application model, according to the corresponding background and characteristics of different projects, carry out data analysis, predict the cause level of the factors, match the data statistical results with the initial network data, and finally calculate the possible costs and benefits of the project by calculating the expected value of the nodes, so as to realize the dynamic technical and economic evaluation of the power grid adapted to different application scenarios.

G. Wang (✉)
State Grid Smart Grid Research Institute CO., LTD, Nanjing 210003, China
e-mail: 26261045@qq.com

State Grid Key Laboratory of Information & Network Security, Nanjing 210003, China

J. Pan · A. Dong
State Grid Jilin Electric Power Company Limited, Changchun 130021, China

C. Lv · B. Zhao
Power Economic Research Institute of Jilin Electric Power Co., Ltd, Changchun 130011, China

S. Patnaik et al. (eds.), *3D Imaging—Multidimensional Signal Processing and Deep Learning*, Smart Innovation, Systems and Technologies 348,
https://doi.org/10.1007/978-981-99-1145-5_7

7.1 Introduction

The digitalization construction of power grid companies will gradually be carried out in depth. At present, there are many researches on digitalization construction, but the research on quantitative control dynamic analysis for the characteristics of the digitalization application scenario of power grid has not been carried out [1, 2]. There is no targeted, quantifiable and systematic benefit evaluation method and dynamic technical and economic evaluation model [3, 4]. There is also no forward-looking research on the technical and economic evaluation that can be quantified for the effectiveness of the digital construction of power grid enterprises to adapt to different application scenarios, and no scientific, complete and quantifiable dynamic evaluation and control method required for the effectiveness evaluation of the digital construction of enterprises has been established.

7.2 Quantitative Control Dynamic Analysis Technique

Quantitative control can greatly reduce the pressure of data processing, which has important research significance both in theory and in practical applications. Since the fifties of last century, some scholars began to study the subject of quantization control in the control field. First, Kalman studied the dynamic performance of the quantized control signal for the sampling system [5]. Since then, the research on quantitative control has become more and more in depth [6]. It is worth pointing out that as a very important research topic, quantitative control has been widely used in practical projects, and has been applied in hybrid systems, digital control systems and information constrained control. As we all know, stability is one of the most important control performance indicators in control systems, and also the basis for achieving other control objectives, which urges further research on quantitative control. In recent years, there have been many quantitative control methods, such as quantitative control methods based on static quantizer and quantitative control methods based on dynamic quantizer [7, 8]. The difference of quantizer and quantization method will lead to the difference of quantization error, and the quantization error, regardless of its size, will have a bad impact on the performance of the system. Therefore, it is important to find a better quantization control method to achieve better performance of the system.

At present, scholars at home and abroad have a strong interest in the research of quantitative control, and have made many meaningful research results in the quantitative control of various control systems. In the aspect of system theory research, reducing quantization error as much as possible can effectively reduce the threat to system stability. Most of the existing research results focus on the stability analysis of the system. There is still much room for the development of quantitative control research for practical application problems, which is of great practical significance

[9, 10]. It is also necessary to improve the system theory while exploring quantitative control methods. Because noise is ubiquitous in industrial production, the quantitative control research in the presence of noise can be considered. In order to improve the performance of the system, the reliability of the system is constrained in advance. In this case, the research on the quantitative control method of the system needs further development and exploration.

7.3 Dynamic Analysis of Quantitative Control of Power Network

The quantitative identification of power grid digitalization is a research method that uses statistical principles and probability theory to explore the quantitative characteristics, quantitative relations, and quantitative changes in the development process of power grid digitalization information, so as to determine the relationship between power grid digitalization information and variables. Quantitative identification is the process of transforming concepts into data and classifying them. Data is used to describe the attribute characteristics and relationship patterns between the digital information of power grid, and accurate qualitative purpose is achieved through quantitative analysis.

The dynamic analysis of quantitative control of power grid adopts such technologies as econometrics, mathematical economics, economic control theory, nonlinear control theory, large system theory, and system dynamics to dynamically adjust the analysis indicators of power grid digital projects according to the characteristics of power grid analysis data, so as to obtain better analysis results. Quantization is to divide the characteristics of the data to be analyzed. Each divided area corresponds to a constant value. The accuracy of the required analysis results is related to the corresponding quantized level (layers). The quantization series and quantization density determine the maximum range of analysis and the best control accuracy that can be obtained.

It is classified according to the subjective and objective characteristics of quantitative identification methods, including qualitative and quantitative categories.

(1) Qualitative method

The meaning of the qualitative method is not to use mathematical methods, but to directly make a qualitative conclusion on the value judgment of the appraisal object based on the appraiser's observation and analysis of the object's usual performance, realistic state or literature.

Qualitative methods are characterized by comprehensiveness, accuracy, and subjectivity. In the real world, some activities are very complex and fuzzy, and there are many factors that are difficult to quantify. Therefore, qualitative evaluation is indispensable.

(2) Quantitative method

The meaning of quantitative method is to use mathematical methods to collect and process data and make a quantitative conclusion on the value of the assessment object.

The characteristics of quantitative methods are objective, clear, and easy to compare. With the development of measurement and quantitative recognition theory, the forms of quantification are more and more, and widely used in many fields.

However, both qualitative and quantitative methods have their advantages and disadvantages, and each has its scope of application. The development trend of modern quantitative identification theory and practice is to combine qualitative and quantitative methods to obtain more objective and comprehensive evaluation results.

7.4 Function Analysis of Power Grid Digitalization Project

The function analysis of power grid digitalization project is to study and analyze the functions of the components of the research object, so as to realize each function of the research object. Accurately and clearly describe the functions of each component of the research object, so that it can clearly define the functions and functions of each component structure. Functional cost method and functional coefficient method are commonly used for functional analysis of power grid digitalization projects. The function cost method is also called the absolute value method, which includes two steps: calculating the function evaluation value and calculating the current cost of the function. Function coefficient method includes calculating function coefficient and calculating cost coefficient.

Three calculation steps for determining the preferred scheme are shown in Fig. 7.1.

(1) Determination of functional coefficient

The scheme function coefficient is the key to calculate the value coefficient, and the determination of the function weight is the key to calculate the function coefficient, which directly determines the result and final judgment of the function coefficient. The methods to determine the functional weight include objective weighting methods such as entropy weight method, multi-objective distance maximum method, and subjective weighting methods such as forced scoring method and analytic hierarchy process. The objective weighting method is calculated based on objective data, which is complicated and the results are easily misled by individual data. The ring

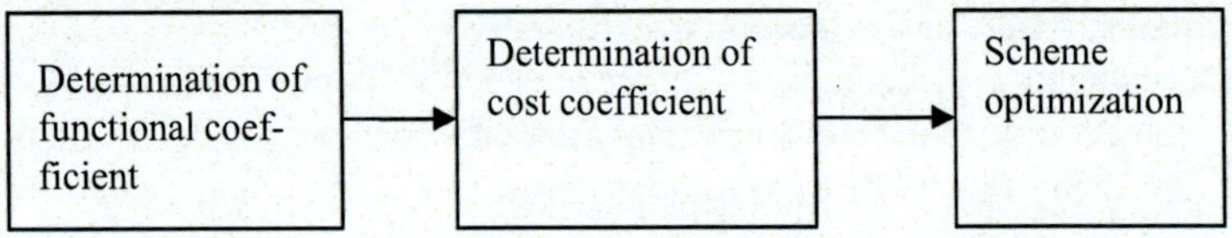

Fig. 7.1 Functional analysis steps of power grid digitalization project

comparison scoring method and forced scoring method in the subjective weighting method are simple, but they basically depend on the level of experts and are subjective. Analytic hierarchy process is a combination of legality and quantification. It is commonly used in determining weights. The calculation process is relatively complex, but it has strong operability. The function system in this paper is determined according to the business scenario, and the objective weighting method is adopted for the relatively fixed digital basic capabilities, common service capabilities, operation support capabilities, and other scenarios of the business scenario. The power grid production, enterprise operation, customer service, industrial ecology, and other scenarios adopt the subjective empowerment method.

(2) Determination of cost coefficient

The cost coefficient is determined by the ratio between the cost of each scheme and the sum of the cost of each scheme.

(3) Scheme optimization

After calculating the function coefficient and cost coefficient of each alternative, the value coefficient of each alternative can be calculated by comparison. The scheme with the largest value coefficient is recommended as the optimal scheme.

In this paper, the value engineering method, expert scoring method, Delphi method, analytic hierarchy process, membership function method, fuzzy comprehensive evaluation method, precedence chart method, cluster analysis method, and other methods are used to realize the dynamic analysis of quantitative control of analysis indicators and weights, economic evaluation models, and technical and economic analysis models.

7.5 Research on Influencing Factor Set of Target Feature Quantification in Digital Application Scene Based on Expert Scoring Method

The construction of influencing factor set for target feature quantification of digital application scenarios based on expert scoring method includes the following three parts, as shown in Fig. 7.2.

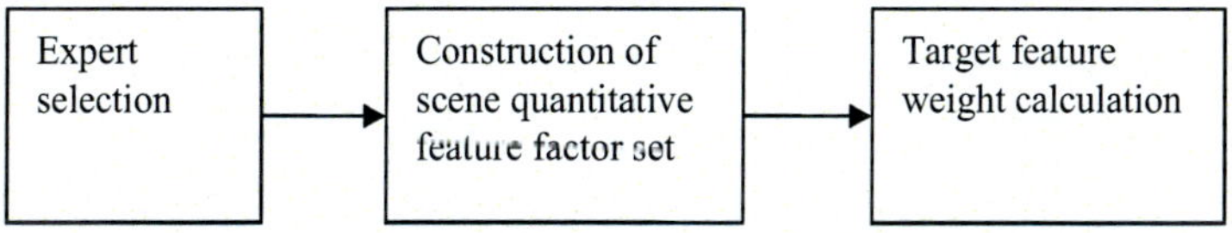

Fig. 7.2 Flow chart of influencing factor set construction

(1) Expert selection

How to select experts is very important when using expert scoring method to determine target characteristics. Experts can be selected from the following aspects:

a. Identification of expert qualification. The selected experts shall be the authorities in the field of digital benefit evaluation research, or experts and scholars with at least rich working experience in this field. The selected experts must be familiar with the characteristics, specifications, and development status of the digitalization of electric power enterprises. To make correct evaluation data, experts need to first understand the technical characteristics of power technology and economic analysis, because the importance of the same indicator in different industries is not the same, even very different; The selected experts should be familiar with the situation of the enterprise, otherwise, the evaluation results are difficult to be practical.
b. Determination of the number of experts. The number of experts should not be too large. On the one hand, with the increase of the number of respondents, the cost of investigation will become higher and higher; On the other hand, the number of people is too large to organize, and the handling of the results is also complicated. However, the number of experts should not be too small. Too few samples will increase the instability of the evaluation results.

(2) Construction of scene quantitative feature factor set

The power grid digital system mainly includes eight types of scenarios, such as digital basic ability, common service ability, operation guarantee ability, power grid production, enterprise operation, customer service, industrial ecology, and government service. According to the specific scenario, the quantitative feature system of digital application scenario is determined from the grid analysis and evaluation index concentration and expert experience.

(3) Target feature weight calculation

Firstly, experts give their normalized score as the weight of the scene features according to the scene features; Next, according to the importance of each scene feature to the scene, four types of evaluation criteria are set for each target feature, namely, very important, important, average, and unimportant. Each expert marks "√" under the criteria that they identify; Then the evaluation statisticians calculate the proportion of experts marking "√" under each type of standard, and use the weighted average method to calculate the score of each target scene feature; Then the score of each target feature is obtained by weighted average of the scores of each target feature and their respective weights, that is, the quantitative value of each target feature.

7.6 Research on Quantitative Impact Index Set of Digital Application Scene Features Based on Fuzzy Analytic Hierarchy Process

The main realization process of using the fuzzy analytic hierarchy process (FAHP) for analysis is: divide the various factors contained in the problem to be studied into several interrelated levels according to a certain membership relationship, and then use specific mathematical methods to calculate and determine the relative importance of the factors at each level and their weight values. Finally, analyze and solve the problem in the order of primary and secondary measurement values from high to low, and draw correct conclusions. The algorithm implementation process is shown in Fig. 7.3.

(1) Target feature system construction

Based on the influence factor set, the expert method is used to construct the target feature system based on the business scenario in combination with the grid digital business scenario.

(2) Determine the weight of each target feature

On the basis of expert consultation and full investigation, we can establish the comparative judgment matrix of the system. Through further calculation, we can get the eigenvalues of the comparison judgment matrix and the corresponding eigenvectors. After the eigenvectors are normalized, we can determine the weight of each target feature in the system. For each target feature, calculate the weight.

(3) Consistency test of comparison judgment matrix

In practice, the comparison judgment matrix we get is not necessarily consistent, that is, it does not meet the transitivity and consistency. Therefore, we should carry out consistency inspection. The consistency test mainly focuses on the following indicators.

(1) Consistency indicators are determined according to grid analysis scenarios.
(2) The random consistency index is determined by expert method.
(3) The consistency ratio indicator is calculated according to the user-defined publicity. When the indicator is < 1, the consistency of the comparison judgment matrix is considered acceptable.

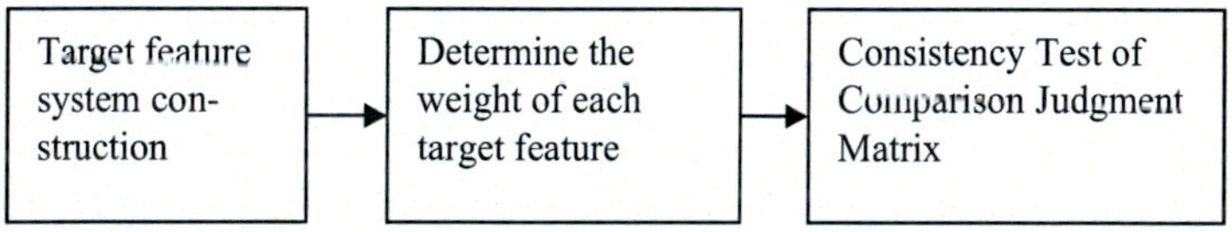

Fig. 7.3 Flow chart of quantitative impact indicator set construction

(4) Establish result comment level

Since the characteristics of each objective of the evaluation system are qualitative descriptions, a rating set of comments should be established for the evaluation results, which correspond to important, general, and unimportant.

(5) Establish decision matrix and quantify target characteristics

After scoring each target feature, the membership of each target feature relative to the comment can be obtained. According to the grid business scenario, the membership coefficient matrix belonging to each target feature is constructed by combining expert method.

(6) Composition operation

Finally, the evaluation index weight vector is multiplied by the membership matrix to obtain the rating vector of each target feature. Determine the weight of the factor set corresponding to each indicator.

7.7 Dynamic Identification Technology of Quantitative Control Based on Bayesian Network

The main process of building the network structure model: first, establish the initial Bayesian network structure diagram through the expert method of fuzzy analytic hierarchy process, and then optimize the network structure through quantitative learning. The Bayesian network model is constructed by combining machine learning with expert experience knowledge. Based on historical experience, we first pre judge the correlation relationship between indicator factors, select the structure algorithm in machine learning to build the initial network structure diagram, and then optimize the initial network structure diagram through the questionnaire survey results of indicator factor correlation. After that, the boundary values of different indicators need to be divided.

(1) Network Structure Learning

The network structure learning adopts K2 structure algorithm and Markov Monte Carlo structure algorithm. The specific construction process is as follows. Firstly, the characteristic quantitative impact index set is used as a priori knowledge, the digital benefit impact factors are input as node variables, the relationship between the various benefit impact factors is preliminarily determined, and finally the Bayesian network structure chart is output.

(2) Bayesian Network Computing

Based on the dynamic identification model of Bayesian network quantitative control, the technical and economic analysis of digital projects is completed.

(1) The expected project costs and benefits are preliminarily calculated using the functional analysis method;
(2) Based on expert scoring method, the quantitative influencing factor set of digital application scene target features is constructed;
(3) Based on fuzzy analytic hierarchy process, the quantitative impact index set of digital application scenario features is constructed;
(4) The application model of index calculation is constructed based on K2 structure algorithm and Markov Monte Carlo structure algorithm;
(5) Dynamic analysis and calculation.

When using the indicator calculation application model to calculate the factors of specific projects, according to the corresponding background and characteristics of different projects, data analysis is carried out, the cause level of factors is predicted, the data statistical results are matched with the initial network data, and the possible costs and benefits of the project are calculated by calculating the expected value of the node.

7.8 Conclusion

In this paper, the dynamic analysis technology of quantitative control of power grid is studied, and the expected project costs and benefits are preliminarily calculated using the functional analysis method; Based on expert scoring method, the quantitative influencing factor set of digital application scene target features is constructed; Based on fuzzy analytic hierarchy process, the quantitative impact index set of digital application scenario features is constructed; Based on K2 structure algorithm and Markov Monte Carlo structure algorithm, the application model of Bayesian network computing is constructed, and a dynamic technical and economic evaluation model adapted to different application scenarios is formed; Based on the calculation application model, according to the corresponding background and characteristics of different projects, carry out data analysis, predict the cause level of the factors, match the data statistical results with the initial network data, and finally calculate the possible costs and benefits of the project by calculating the expected value of the nodes, so as to realize the dynamic technical and economic evaluation of the power grid adapted to different application scenarios.

Acknowledgements This work is supported by Science and Technology Project of State Grid Corporation of China (Research on the evaluation system and technology tools of digital technology and economy of enterprises, No. 5700-202129180A-0-0-00).

References

1. Xiang, M., Chenhan, S.: Practice of enterprise architecture method based on TOGAF. Digit. Commun. World **4**, 194–196 (2017)
2. Yoo, Y., Henfridsson, O., Lyytinen, K.: Research commentary—the new organizing logic of digital innovation: an agenda for information systems research. Inf. Syst. Res. **21**(4), 724–735 (2010)
3. Bogner, E., Voelklein, T., Schroedel, O.: Study based analysis on the current digitalization degree in the manufacturing industry in Germany. Procedia CIRP **57**, 14–19 (2016)
4. Jun, W., Yongqiang, H.: Discussion on information architecture management of power enterprises. Power Inf. Commun. Technol. **14**(6), 14–17 (2016)
5. Paulus, R.D., Schatton, H., Bauernhansl, T.: Ecosystems, strategy and business models in the age of digitization-how the manufacturing industry is going to change its logic. Procedia CIRP **57**, 8–13 (2016)
6. Bin, C., Xiaoyi, Y.: Research and practice of power enterprise information management based on enterprise architecture. Enterp. Manage. **S2**, 526–527 (2016)
7. Tong, Y., Li, Y.: The Evaluation of enterprise informatization performance based on AHP/GHE/DEA. In: Proceeding of the International Conference on Management
8. Jiachen, T.: Research on enterprise information architecture design of the company. Wuhan University of technology (2016)
9. Yongwei, L., Qian, S., Bo, L.: Enterprise information architecture design. China Sci. Technol. Inf. **18**, 43–44 (2015)
10. Li, Z.: Research and application of digital transformation technology in power enterprises. China New Commun. **01**, 127–128 (2013)

Chapter 8
Research on Detection of Fungus Image Based on Graying

Chengcheng Wang and Hong Liu

Abstract With the improvement of people's living standards, people pay more and more attention to the harm of fungi. Inspired by the smartphone-based aflatoxin rapid detection system proposed by Zhang Liyong, this paper improves the method to a certain extent. First, the average value of the blue component was added to Zhang's dynamic method to reduce the effect of excess blue component caused by UV light exposure, which was found to mark the white light in the picture. Therefore, the average value of red is introduced to exclude the influence of white light. When this method is applied to single chip microcomputer, it is found that the image produced by single chip microcomputer has great noise. Therefore, median filtering is added to eliminate noise, so as to achieve the purpose of single chip identification.

8.1 Introduction

With people living standard improving, people pay more and more concerned with fungal harm. According to the National Institute of Health and Medicine and Medical Journal: Their researchers analyzed the case files of 1,100 adult asthma patients and found that patients who were allergic to mold were five times higher than those who were allergic to other substances [1]. Fungus can also hide a large number of viruses such as hepatitis virus, so experts remind the public to often pay attention to home fungus. In fact, fungus are everywhere in life. There are plenty of fungus hidden in nail seams, feet, sweaty skin folds, and even acidic digestive tracts. Harm of mold pollution to human body: the toxin produced by mold will cause nervous and endocrine disorders, immunosuppression, carcinogenesis, teratogenesis, liver and kidney damage, reproductive disorders, and other diseases to people and livestock. If people stay in the indoor environment where the wall is polluted by mold for a long time, they will show discomfort, most of which are headache, chest tightness, rhinitis, pharyngitis, easy fatigue, irritability, skin allergic reaction, and other discomfort [2].

C. Wang (✉) · H. Liu
Shanghai DianJi University, Shanghai, China
e-mail: 2474273800@qq.com

S. Patnaik et al. (eds.), *3D Imaging—Multidimensional Signal Processing and Deep Learning*, Smart Innovation, Systems and Technologies 348,
https://doi.org/10.1007/978-981-99-1145-5_8

However, most of the methods adopt the method of desampling for detection by means of chemistry or sampling.

Zhang [1] **proposed a smartphone-based rapid detection system for aflatoxins.** Because it only detects the mold in the picture through a simple method, taking into account the impact of excessive blue light components caused by ultraviolet lamp irradiation, making the detection inaccurate. Based on the above research background, this paper attempts to improve Zhang Liyong's scheme to make it more accurate in recognition, and tries to carry its method on the single chip microcomputer.

8.2 Fungus Image Gray Processing

8.2.1 Graying of Fungus Pictures

In the current common image storage formats, 24-bit color images are used. In a 24-bit color image, each pixel is represented by three bytes, typically Red, Green, Blue (RGB), the three primary colors of light. A threshold is selected to determine that each color light corresponds to an eight-bit binary (0–255) to represent the intensity of the color light, 0 is the weakest and 255 is the strongest [3]. $R = G = B = 0$ is black, $R = G = B = 0$ is white. Is black and fits the color we need to filter. Therefore, it is necessary to convert a color image into a grayscale image, that is, each pixel of the color image of the original RGB through a suitable algorithm to get a gray value, as the pixel gray value. And a certain processing, and then by these pixels re-combination to get a gray image. The commonly used gray value calculation method is as follows:

1. Component method: choose a color value as the gray value.
2. Maximum method: select the largest RGB value as the gray value of the point.
3. Mean method: select the average RGB three colors as the gray value of the point [4].
4. Weighting method: Due to the different sensitivity of human eyes to color, the following gray value formula can be obtained according to the sensitivity of human body to color.

$$\text{Gray} = 0.21\text{R} + 0.71\text{G} + 0.07\text{B}$$

Under the ultraviolet lamp, the fluorescence reaction of aflatoxin b in mold occurs, and there should be two types of fluorescence reaction. Type B is purple fluorescence, while type G is green fluorescence under the ultraviolet lamp. Therefore, the fluorescence part of the picture needs to be enhanced to make the feature more obvious. Because the fluorescence color of the mold is only green and purple fluorescence and its color is monochromatic. Therefore, directly analyzing the available information of the color in the picture can obtain its aflatoxin. Therefore, directly analyzing the

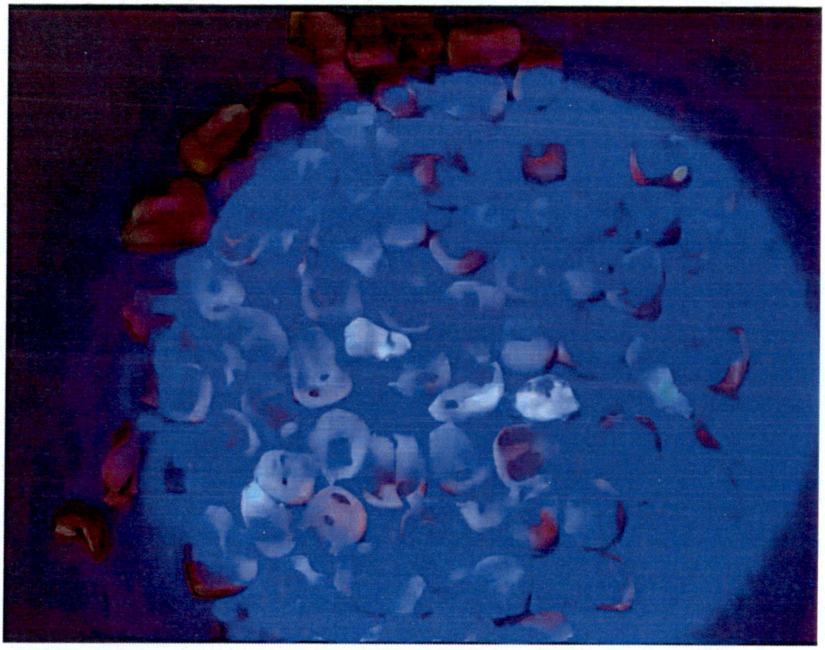

Fig. 8.1 Original

content of the two colors in the photo can directly reflect the two aflatoxins in the sample.

The above four algorithms are tested, respectively. Below is the image of peanut under UV light, where the blue accentuated fluorescence of aflatoxin is shown in Fig. 8.1. The image was calculated using the above method to obtain the corresponding gray value.

The following is the calculated average gray value. Component method: 52.69387739087589 (blue).

8.254235806094716 (green).

23.12091861557056 (red).

Maximum value method: 52.96234914.

Method of mean value: 27.9142402.

Weighted method: 4.54012497.

It can be seen that the results of the maximum value method and the component method (blue) are basically the same, while the average gray value obtained by the mean value method and the weighting method is too small. The information obtained by the component method can only represent the information in the blue part, but not the green part. So the maximum value method is too large, it cannot represent the two types of information. Using the combination of weighting method and component method, the gray value will change with the blue gray value and green gray value of the pixel. Because the total gray value is too small [5]. The average value of one of the variables is too small. Assuming that the ratio of the blue-green brightness component in one measurement is x: y, the gray value can be expressed as a formula:

$$\text{Gray} = \frac{xB + yG}{x + y} \tag{8.1}$$

More blue due to the role of ultraviolet lamp, so deform this dynamic formula and introduce the blue gray of mean. Use the mean to reduce too much blue results are as follows:

$$\text{Gray} = \frac{x(B - \overline{x}) + yG}{x + y} \tag{8.2}$$

At the same time, there is a case where B is less than the mean, so a combination of the two is used. B is less than the blue gray mean, then select the undeformed formula. When B is greater than the blue gray, select the deformed gray formula.

8.2.2 *Threshold Method*

Because the entire image gets too much information and more than is needed, it is impossible to enhance it all. Enhancement using threshold method, not all pixels are enhanced. When the gray value of a pixel is higher than this threshold can be considered as a valid point, otherwise it is considered to be no sample space. After testing different thresholds, you can get Fig. 8.2, where the cumulative number of points of the ordinate sampling, and the abscissa is the threshold of the sampling, where the threshold is 255.

It can be seen from Fig. 8.3 that the threshold has an inflection point at about 45, and the image at the threshold of 45 can be obtained as 3 (where the marking part is directly whitened to facilitate recognition), and the image is just in the ultraviolet lamp. It cannot identify the results. There is an inflection point at 145 of the image, and the image with a threshold of 135 can be obtained 4.

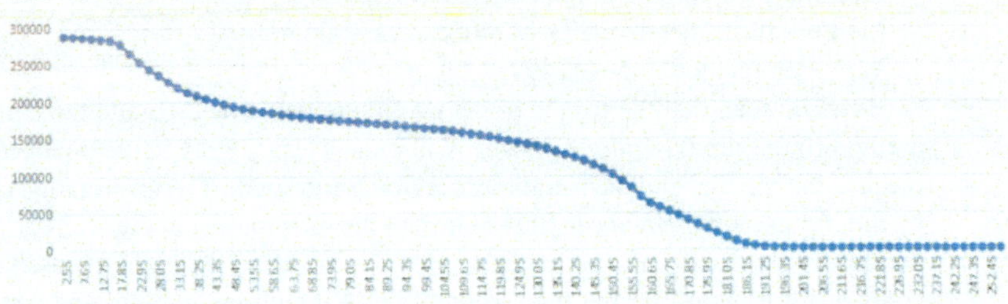

Fig. 8.2 Scatter plot

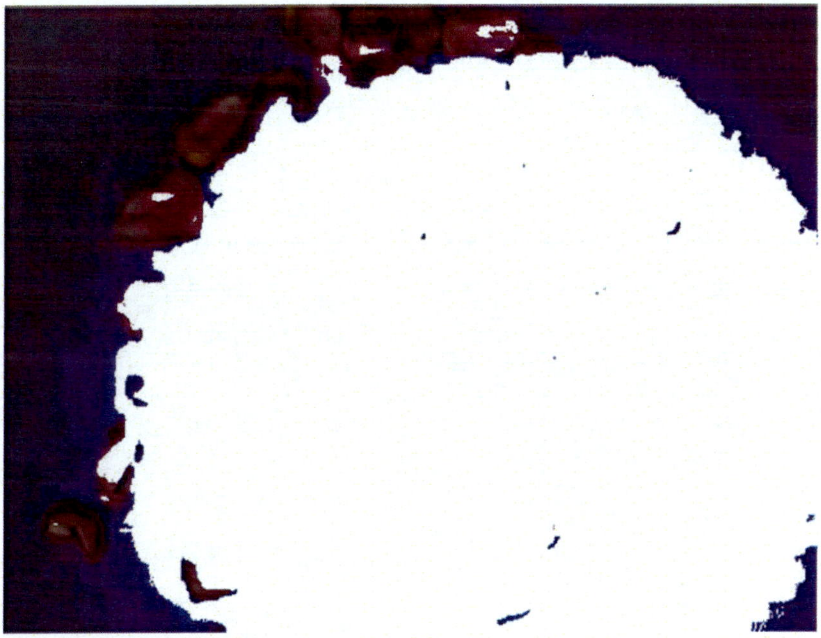

Fig. 8.3 Figure with value of 45

Obviously, Fig. 8.4 is the information we need. In order to make it not limited to only one picture effective and more suitable for other pictures, a dynamic threshold method is proposed. $\overline{x}$ is blue light. From the formula can be obtained its coefficient and the relationship between the available points Fig. 8.5.

$$yu = \overline{x} * \alpha$$

From the image, it can be concluded that the coefficient is around 3.0, and the inflection point is obvious. Call out the image to get Fig. 8.6 (where the white part is the marker).

Annotation from Graph Analysis Part is the brightness required.

8.2.3 *Problems with Testing*

In order to make the image more available, a variety of pictures are used for testing, and it is found that the recognition is easy to mark the picture itself white (Fig. 8.7). So remove the white effect.

This article uses a similar approach to Sect. 1.2. Because white has a red component. Therefore, add red screening. The screening method is that the multiple of the

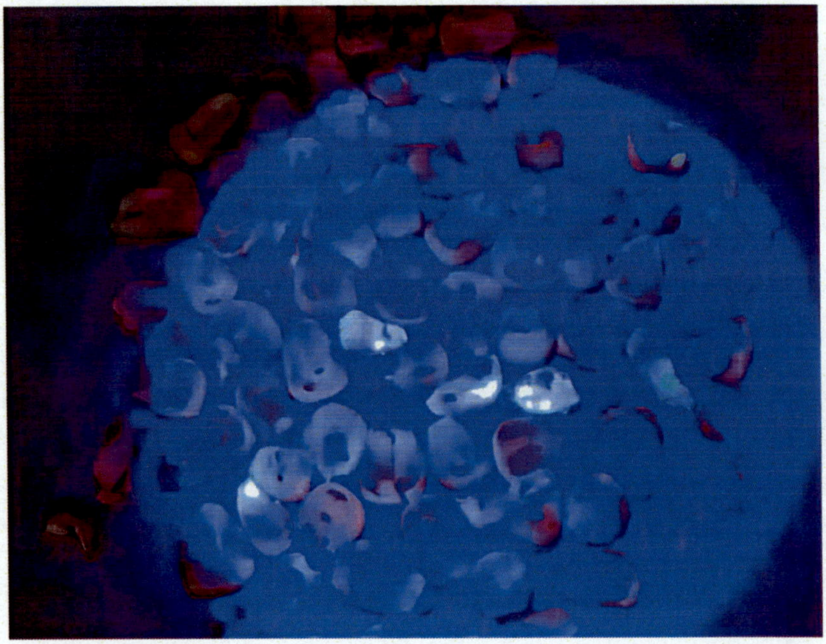

Fig. 8.4 Graph with value 135

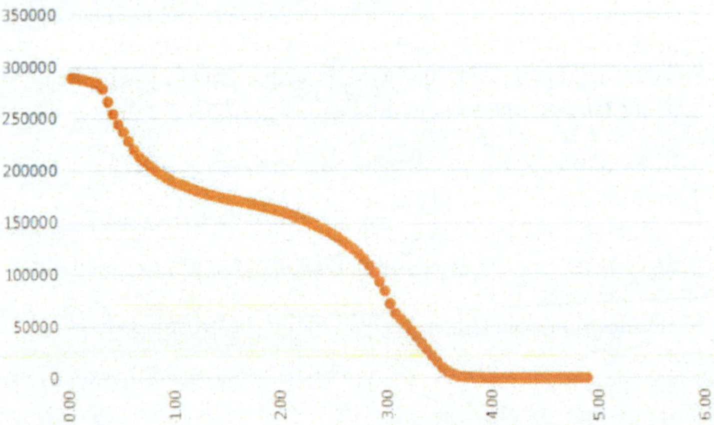

Fig. 8.5 Scatter plot

gray value of the red part of the point is less than the average blue gray value, so that the point is the part illuminated by the ultraviolet lamp. Not the white light part. The different processing results are shown in Fig. 8.8.

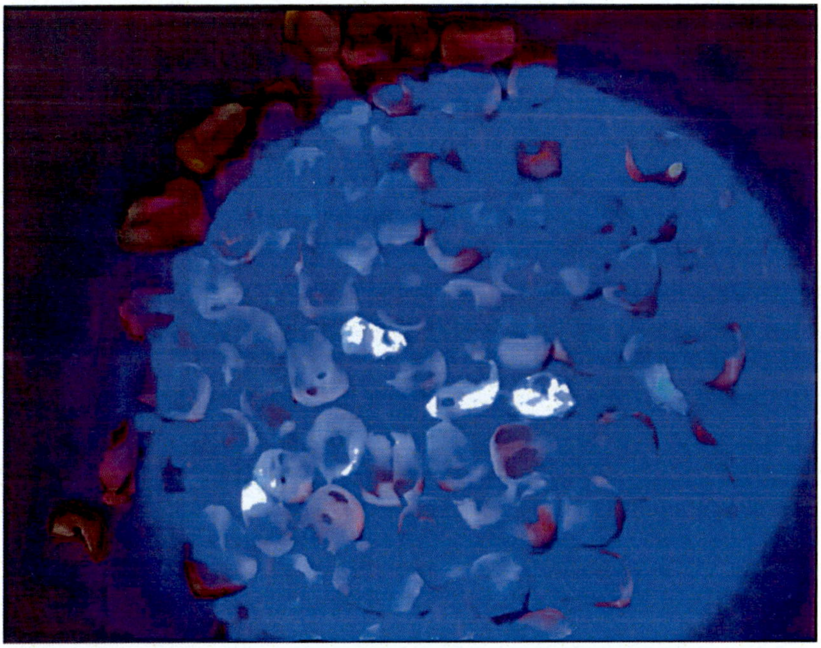

Fig. 8.6 Final drawing

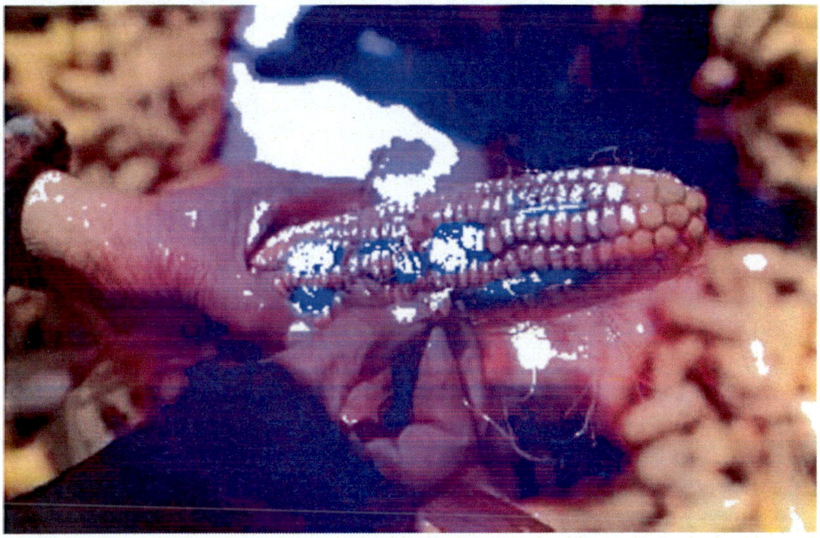

Fig. 8.7 Test picture results

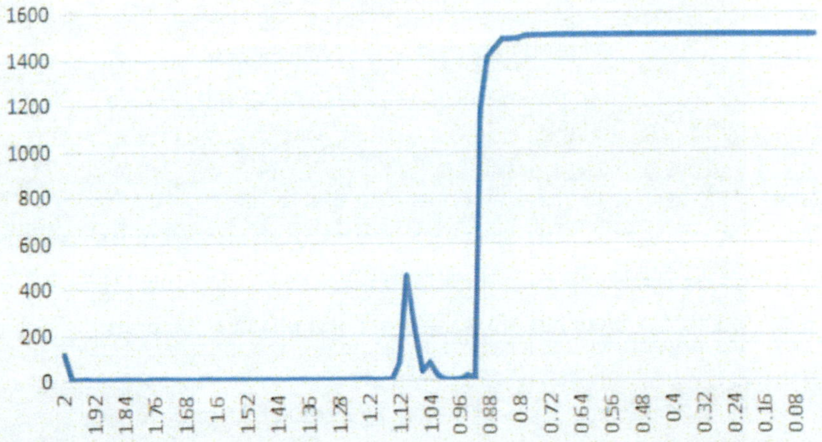

Fig. 8.8 Diagram

By observing the image, it can be seen that its 1.12 is the fluctuation point, and it can be seen in the 1.12 image. It can be seen from Fig. 8.9 (the white part is the identified mold part), which can basically screen out the ultraviolet irradiation part.

Fig. 8.9 Improved figure

8.3 Realization of Single Chip Microcomputer

8.3.1 Selection of Single Chip Microcomputer

This article chose the cheaper ESP32-CAM. ESP32-CAM is the latest small-size camera module released by Anxinke. The module can work independently as a minimum system, with a size of only 27 * 40.5 * 4.5 mm and a deep sleep current of 6 mA. Camera module Select OV2640 module.

8.3.2 Total Process of Single Chip Microcomputer

The process implemented by the single chip microcomputer in this paper is to initialize the camera and the Wi-Fi module and obtain the image of the camera, and then perform filtering on the image to reduce the impact of image noise on recognition. Finally upload in Bemfa Cloud. Specific flow chart is shown in Fig. 8.10.

8.3.3 Selection of Filter

Considering the performance of the single chip, we only choose the low-pass filter. The output of the smoothed linear spatial filter is a simple average of the pixels contained in the neighborhood of the corresponding pixel points, the averaging filter. The mean filter is also a low-pass filter [6]. The mean filter is easy to understand, that is, the average value in the neighborhood is assigned to the central element. Taking the 3 * 3 mean filter as an example, the principle of the mean filter algorithm is as follows: Fig. 8.9 and compare Fig. 8.11 (Fig. 8.12).

It is found that the filtered noise is not completely.

The principle of median filtering is to use the median value of the pixels around the point to make the pixel value closer to the real pixel value when calculating the modified pixel point, so as to achieve the purpose of eliminating separate noise points [7]. Taking 3 * 3 median filter as an example, the principle of median filter is shown in Fig. 8.13 and compared with that in Fig. 8.14.

Figure 8.14 surface 3 * 3 mean method removes a lot of noise, so this paper uses 5 * 5 mean method for further filtering. The principle of the median filter algorithm is as follows compare Fig. 8.15 [8].

Figure 8.15 Surface 5 * 5 mean method has basically removed the noise.

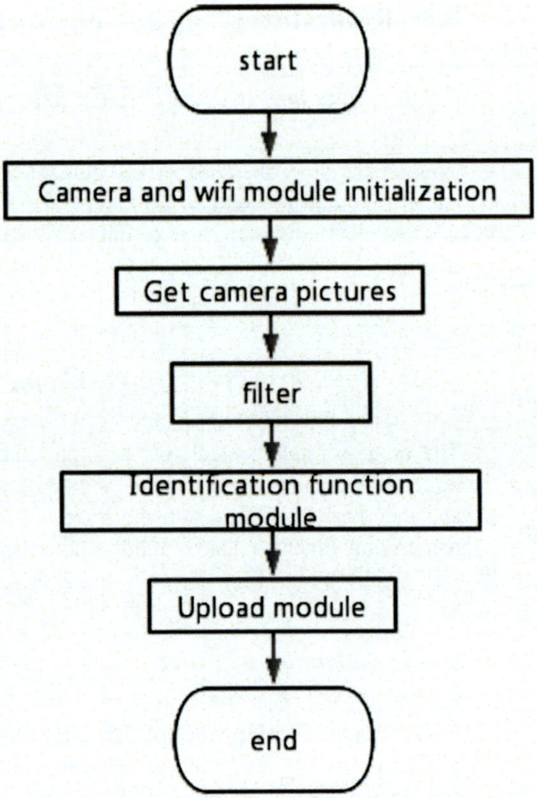

Fig. 8.10 Main flow chart

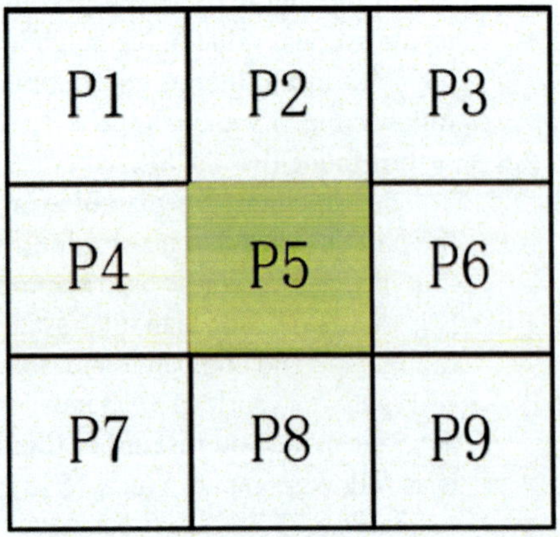

Fig. 8.11 Schematic diagram of mean filtering

$$P_5 = \frac{1}{9} * \sum_{i=1}^{9} P_i$$

↑

Fig. 8.12 Comparison chart of mean value filtering

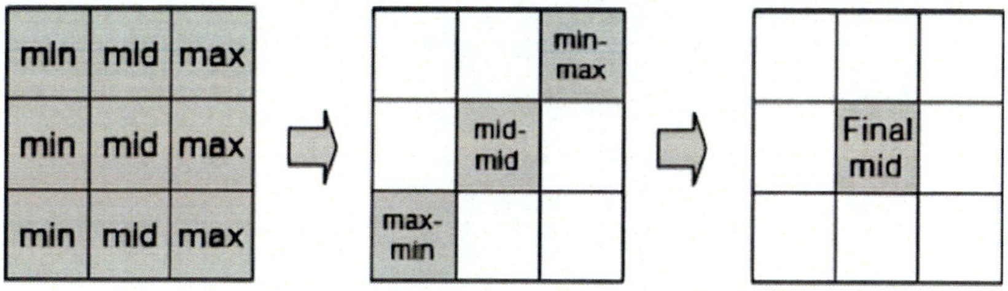

Fig. 8.13 Schematic diagram of median filtering

8.3.4 Detection Function Module

Testing with the method in Chap. 1 yields Fig. 8.16.

It can be seen from the test results that the recognition can be basically realized when the camera is out of the original color, and the progress test needs to change the equipment for the test. We collected 100 fluorescent pictures of bananas under different conditions through reptiles and other means. The system was used for detection, respectively, and it was found that it can only have a good effect in the ultraviolet lamp 365 mm, and 70 of them can detect the obvious fluorescence effect. And some weak fluorescence effects were not detected. It has a better effect. The effect of other wavelengths is not ideal [8, 9].

Fig. 8.14 Comparison chart of median filtering

Fig. 8.15 Comparison chart of 5 * 5 median filtering

Fig. 8.16 Actual test chart

8.4 Summary

This paper improves Zhang Liyong's mold detection, which can basically complete the detection of some purple light molds, but cannot detect the identification under strong sunlight or other unconventional conditions. The single chip implementation because of the choice of camera pixel is too low, and the limited performance of single chip. So the actual recognition effect is not good, to be further optimized. At present, we plan to use the target detection model in depth learning to detect in the next step.

References

1. Zhang, L.Y.: Rapid detection of aflatoxins based on smartphones. University of Electronic Science and Technology of China (2019)
2. Zhang, X.P.: Study on the distribution of TCM syndromes of unstable angina pectoris with anxiety and depression. Henan University of Traditional Chinese Medicine (2019)
3. Chen, H.: Variation reduction in quality of an optical triangulation system employed for underwater range finding. Ocean Eng. **29**(15), 1871–1893 (2002)
4. Wang, Ya, Zhang, B.F.: Infrared forest fire monitoring system based on saliency detection. Fire Sci. Technol. **037**(012), 1700–1703 (2018)

5. Gadi, V.K., Alybaev, D., Raj, P., Garg, A., Mei, G., Sreedeep, S., Sahoo, L.: A Novel python program to automate soil colour analysis and interpret surface moisture content. Int. J. Geosynth. Ground Eng. **6**(2), 21.1–21.8 (2020)
6. Yu, L., Shi, Z., Fang, C., et al.: Disposable lateral flow-through strip for smartphone-camera to quantitatively detect alkaline phosphatase activity in milk. Biosens. Bioelectron. **69**, 307–315 (2015)
7. Qu, W., Sun, X.X.: Design and hardware implementation of image compression denoising based on median filter and wavelet transform, Chongqing, China. Trans Tech Publications Ltd (2014)
8. Wu, X.J., Zuo, G.K., Yang, Z.Z., et al.: Research on a denoising method of the force sensor signal based on multiple filter. In: Frontiers of mechanical engineering and materials engineering III: selected, peer reviewed papers from the 2014 3rd International Conference on Frontiers of Mechanical Engineering and Materials Engineering (MEME 2014), 21–23 November 2014, pp. 613–618. Trans Tech Publications, Xiamen, China (2014)
9. Wen, S.: Translation analysis of English address image recognition based on image recognition. EURASIP J. Image Video Process. **1**, 1–9 (2019)
10. Rubinstein, C., Limb, J.: On the design of quantizers for DPCM coders: influence of the subjective testing methodology. IEEE Trans. Commun. **26**(5), 565–572 (1978)
11. Li, H.R.: Distributed passive sensor information fusion technology based on neural network. Army J. **041**(001), 95–101 (2020)

Chapter 9
Secondary Frequency Regulation Control Strategy of Battery Energy Storage with Improved Consensus Algorithm

Linlin Hu

Abstract In order to improve the frequency stability of the microgrid, this paper proposes a two-layer strategy for secondary frequency modulation of battery energy storage based on an improved consensus algorithm. The control strategy firstly constructs the objective function of secondary frequency modulation. Secondly, in order to solve objective function, it proposes the frequency response consistency iterative calculation method, so as to achieve effective frequency allocation and rapid recovery. Simulation experiments show that this strategy can quickly restore frequency stability under system imbalance and has faster dynamic response speed and lower overshoot.

9.1 Introduction

To improve the dynamic power balance of load side and generation side in active distribution network, reference [1] introduces consistency algorithm into frequency modulation control method to meet the power balance constraint in economic dispatching. The traditional maximum power point tracking algorithm is the basic control characteristic of wind turbine. Reference [2] carries out inertia frequency modulation coordinated control on the traditional wind turbine and combines the maximum power point tracking algorithm to realize the primary frequency modulation control; this control algorithm reduces the capacity of energy storage unit.

Reference [3] considers battery energy storage systems with different battery state of charge (SOC) constraints, which can realize the function of battery energy storage system to stabilize the frequency fluctuation and further maintain the frequency stability of power grid. In the control process of microgrid inverter, power grid frequency fluctuation will be caused by load fluctuation, parameter change, and

L. Hu (✉)
College of Electrical and Electronic Engineering, Guangdong Technology College, Zhaoqing, China
e-mail: hulinlin1201@163.com

S. Patnaik et al. (eds.), *3D Imaging—Multidimensional Signal Processing and Deep Learning*, Smart Innovation, Systems and Technologies 348,
https://doi.org/10.1007/978-981-99-1145-5_9

nonlinear characteristics. Reference [4] has carried out research in combination with virtual synchronous machine control and proposed an adaptive anti-robust control method, which has better dynamic response speed and can realize the function of rapid error recovery. Making full use of energy storage system is the technical development trend at this stage [5]. Reference [6] proposed a power grid frequency modulation method considering dynamic SOC base point and sag coefficient, which can ensure long-term stable control of frequency. References [7, 8] introduced the concept of interface agent based on the network structure and communication mode of distribution network and proposed a coordinated and optimized operation method of active distribution network based on consistency algorithm.

This paper proposes a two-layer strategy of secondary frequency modulation based on improved consistency algorithm. It first constructs the secondary frequency modulation objective function and sets an effective battery energy storage output structure. Secondly, the frequency response consistency iteration method is used to solve the objective function, to realize the effective frequency allocation and rapid recovery of the upper and lower control. Finally, the simulation experiment of the control strategy under actual working conditions is carried out through MATLAB/Simulink. This strategy can quickly restore frequency stability under system imbalance and has faster dynamic response speed and lower overshoot.

9.2 Optimal Control Method of Secondary Frequency

9.2.1 Energy Storage Output Control Structure

Both the rapid recovery of battery energy storage and the power grid frequency modulation need to set a reasonable control law of battery energy storage output, which not only needs to meet the demand of battery energy storage capacity, but also can improve the power grid frequency modulation effect. Based on the limitations and requirements of the power grid and the energy storage, it comprehensively sets the energy storage output control objectives and establishes two control methods of fast recovery and fast frequency modulation.

9.2.2 Secondary Frequency Modulation Objective Function of Power Grid

The setting of cost objective function is shown in Eqs. (9.1)–(9.5). Opt is the optimal objective function, and $\mathrm{Cost}_{\mathrm{load}}$, $\mathrm{Cost}_{\mathrm{gen}}$, and $\mathrm{Cost}_{\mathrm{battery}}$ are the secondary frequency modulation costs of flexible load unit, power supply unit, and battery energy storage unit.

$$\mathrm{Opt(Cost)} = \mathrm{Cost}_{\mathrm{gen}} + \mathrm{Cost}_{\mathrm{load}} + \mathrm{Cost}_{\mathrm{battery}} \tag{9.1}$$

$$\mathrm{Cost}_{\mathrm{battery}} = \sum_{1}^{N} \omega_{\mathrm{EMS}-c}\left(\omega_{\mathrm{EMS}-f} \times \Delta f\right) \tag{9.2}$$

$$\mathrm{Cost}_{\mathrm{load}} = \sum_{1}^{N} \left(D_{1j} \times \left(P_L^j\right)^2 + E_{1j} \times P_L^j + F_{1j}\right) \tag{9.3}$$

$$\mathrm{Cost}_{\mathrm{gen}} = \sum_{1}^{N} \omega_{\mathrm{gen}-c}\left(\omega_{\mathrm{gen}-f} \times \Delta f\right) \tag{9.4}$$

$$P_L^j - P_{L0}^j = \omega_{\mathrm{load}-f} \times \Delta f \tag{9.5}$$

where $\omega_{\mathrm{EMS}-c}$ and $\omega_{\mathrm{EMS}-f}$ are the secondary frequency modulation cost coefficient and response coefficient of battery energy storage unit, respectively, $\omega_{\mathrm{gen}-f}$ and $\omega_{\mathrm{Gen}-c}$ are the secondary frequency modulation cost coefficient and response coefficient of power supply unit, respectively. D_{1j}, E_{1j}, and F_{1j} are the cost coefficients of the flexible load unit, respectively, $P^j{}_L$ and $P^j{}_{L0}$ are the operating power and initial operating power of the flexible load unit, and $\omega_{\mathrm{load}-f}$ is the secondary frequency modulation response coefficient of flexible load unit.

In addition, the boundary conditions of the objective function include power balance constraints and output constraints of the frequency modulation unit, respectively, and its discrete form is:

$$0.25 p_{\mathrm{gen}}^{\max} \geq \omega_{\mathrm{gen}-c} \Delta f \geq 0 \tag{9.6}$$

$$p_{\mathrm{EMS}}^{\max} \geq \omega_{\mathrm{EMS}-f} \Delta f \geq p_{\mathrm{EMS}}^{\min} \tag{9.7}$$

$$P_L^{\min} \geq \omega_{\mathrm{load}-f} \times \Delta f + P_{L0}^j \geq P_L^{\max} \tag{9.8}$$

$$\omega = \sum_{1}^{N} \omega_{\mathrm{gen}-c} + \sum_{1}^{N} \omega_{\mathrm{EMS}-c} + \sum_{1}^{N} \omega_{\mathrm{load}-c} \tag{9.9}$$

In this paper, the control objective function and boundary conditions are Lagrangian functions, and its partial derivatives are solved.

$$\mathrm{load} = \mathrm{Cost}_{\mathrm{gen}} + \mathrm{Cost}_{\mathrm{load}} + \mathrm{Cost}_{\mathrm{battery}} + \theta\left(P - P_f\right) \tag{9.10}$$

where Δf is the change of frequency, $p_{\mathrm{gen}}^{\max}$ is the maximum power of power supply unit, $p_{\mathrm{EMS}}^{\max}$ is the maximum power of battery energy storage unit, ω is the response factor, load is the cost, and θ is the angle.

9.3 Secondary Frequency Modulation Based on Consistency Algorithm

9.3.1 Iterative Calculation Method of Frequency Response Consistency

The frequency response consistency iteration algorithm can realize the cluster division of different distributed frequency modulation units in the microgrid and can gradually divide the energy storage unit, generation side unit, and flexible load unit, respectively. The consistency indicators of different frequency modulation cluster units need to interact, and the same type of cluster units can establish communication links to further realize the demand response of the system. The consistent iterative algorithm can reflect the adjustment differences of different types of adjustment resources and can distinguish different frequency modulation units while ensuring that the demand response coefficient meets the requirements, which effectively improves the calculation speed of the iterative algorithm.

9.3.2 Double-Layer Cooperative Control of Secondary Frequency Modulation for Battery Energy Storage

Based on the technical characteristics of battery energy storage unit, this paper proposes a two-layer coordinated control strategy for secondary frequency modulation of battery energy storage based on consistency algorithm. The lower control strategy takes the optimal cost of frequency modulation of battery energy storage unit as the objective function, and the upper control strategy takes the optimal cost of frequency modulation of flexible load unit and power supply unit as the objective function. The core control method of double-layer control is to adopt consistency algorithm to realize the distributed control of frequency modulation units such as flexible load unit, power supply unit, and battery energy storage unit, which is conducive to rapid recovery of system frequency deviation. Through the coordinated control of the upper and lower layers, the control strategy can realize the coordinated and optimized operation of frequency modulation units such as flexible load unit, power supply unit, and battery energy storage unit.

9.4 Simulation Verification

To verify the effectiveness and feasibility of the strategy, it builds the power grid frequency response model on MATLAB/Simulink simulation. This model includes hydraulic generator unit, thermal generator unit, flexible load unit, and battery energy

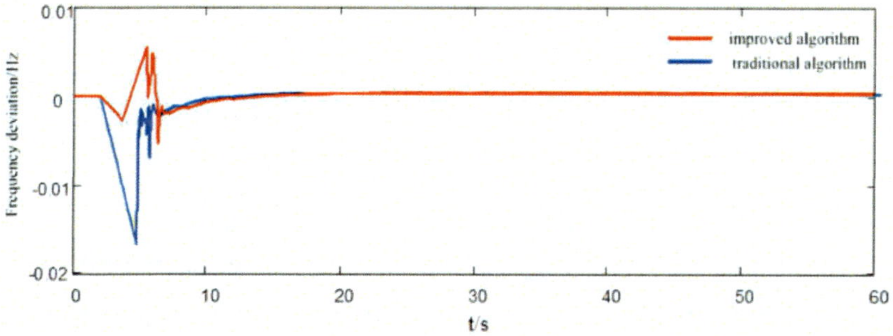

Fig. 9.1 Comparison curve of system frequency deviation

storage unit. Among them, the hydraulic power generation unit is mainly composed of a hydropower unit, and its rated power is 10 MW. The thermal power generating unit is mainly composed of a thermal power unit, and its rated power is 7 MW. The flexible load unit is composed of three groups of AC and DC loads, and its rated power is 15 MW. The battery energy storage unit is composed of two sets of battery energy storage systems. The battery state of charge (SOC) of the battery system is set to 0.7 and 0.8, respectively, and the power cost coefficient is 0.75 and 0.65, respectively. This paper will verify the control strategy under different actual working conditions and test the superiority by comparing different frequency modulation algorithms.

This paper simulates the simulation experiment under the step change of load. At 5 s, the system adds step load fluctuation to verify the anti-disturbance ability. Figure 9.1 shows the comparison curve of the system frequency deviation. Due to the step disturbance of the system load, the system frequency begins to drop sharply, and the frequency drop of the traditional algorithm [6] is large, while the frequency drop of the improved algorithm is smaller than that of the traditional algorithm.

Figure 9.2 shows the comparison SOC curve of the energy storage battery unit under the load step fluctuation. In the simulation experiment, the initial SOC is set to 80%. In the simulation process, the system applies the load step disturbance in 0.5 s, and the SOC begins to decline, which is the discharge process of the battery unit. In the initial stage, the energy storage SOC controlled by the improved secondary frequency modulation algorithm decreases faster. When it is close to 10 s, the energy storage SOC controlled by the traditional secondary frequency modulation algorithm decreases faster. In different periods of SOC decline, the SOC of energy storage under the two control methods decreases at a constant rate. In the 10–20 s stage, the improved secondary frequency modulation algorithm can quickly achieve the SOC stability, has a good adjustment time, and will not produce large fluctuations in the system frequency.

Figure 9.3 shows the comparison curve of output power of energy storage unit under load step fluctuation. It can be seen from Fig. 9.2 that the battery energy storage unit can automatically adjust the charging and discharging power of the energy storage unit under the influence of SOC and generator set output power changes. The larger the regulating power of the energy storage unit, the more effective the

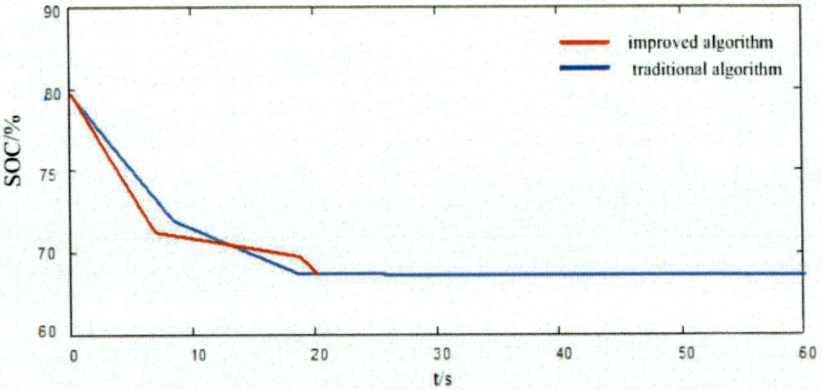

Fig. 9.2 SOC comparison curve of energy storage unit

system frequency deviation caused by load imbalance can be balanced. The output power of the battery unit controlled by the traditional algorithm has a large overshoot.

Figure 9.4 shows the output power comparison curve of the generator set under the load step fluctuation. Two control strategies can maintain the power output stability of the generator set. However, the improved control strategy can maintain the output power stability in 15 s, but the traditional control strategy needs to maintain the output power stability in 18 s. And the overshoot of the power generation unit under the traditional control strategy is higher than that under the improved control strategy.

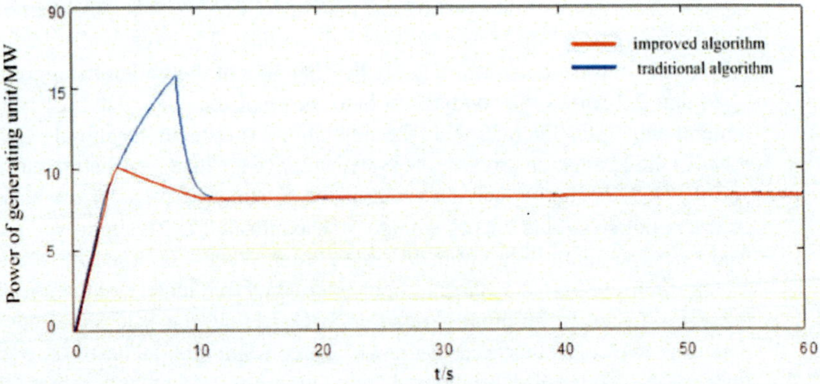

Fig. 9.3 Comparison curve of output power of generating unit

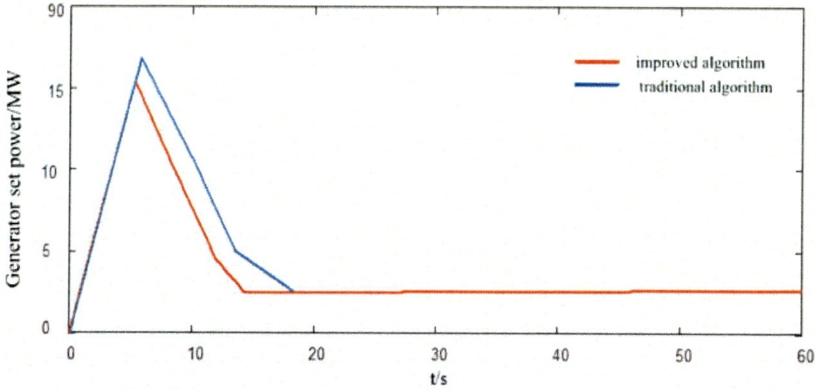

Fig. 9.4 Comparison curve of generator set output power

9.5 Conclusions

In view of the poor stability and anti-interference ability of the secondary frequency modulation of the power grid, this paper proposes a double-layer control strategy of secondary frequency modulation of battery energy storage based on improved consistency algorithm. Firstly, it constructs the secondary frequency modulation objective function and sets an effective battery energy storage output structure. Secondly, the frequency response consistency iteration method is used to solve the objective function, so as to realize the effective frequency allocation and rapid recovery of the upper and lower control. Simulation results show that the proposed control strategy can quickly restore the frequency stability under system imbalance, and not only has faster dynamic response speed, but also has lower overshoot.

Acknowledgements This paper is supported by the 2020 Guangdong Provincial Department of education young innovative talents project in Colleges and Universities (2020KQNCX113); "Key cultivation project" of Guangdong Technology College in 2021(2021GKJZD001).

References

1. Bian, X.Y., Sun, M.Q., Zhao, J., Lin, S.F., Zhou, B., Li, D.D.: Distributed coordinative optimal dispatch and control of source and load based on consensus algorithm. Proc. CSEE **41**(04), 1334–1347+1540 (2021)
2. Yan, X.W., Wang, D.S., Yang, L.L., Jia, J.X., Li, T.C.: Coordinated control strategy of inertia support and primary frequency regulation of PMSG. Trans. China Electrotech. Soc. **36**(15), 3282–3292 (2021)
3. Jia, X.C., Li, X.J., Wan, J., Li, W.Q., Huo, F.Q.: Control method of large-scale battery energy storage system for suppressing the disturbance of power grid. Electr. Power Constr. **41**(06), 69–76 (2020)

4. Zhang, Y.H., Huang, K., Wang, Z.N., Sun, X.P.: VSG adaptive backstepping robust control strategy of VSG based on secondary frequency modulation. Power Syst. Technol. **45**(05), 1985–1992 (2021)
5. Li, J.H., Gao, Z., Ying, H., Lin, L., Shen, B.X., Fan, X.K.: Primary frequency regulation control strategy of energy storage based on dynamic droop coefficient and SOC reference. Power Syst. Prot. Control **49**(05), 1–10 (2021)
6. Yan, G.G., Liu, Y., Duan, S.M., Li, H.B., Mu, G., Li, Ju.H.: Power distribution strategy for battery energy storage unit group participating in secondary frequency regulation of power system. Autom. Electr. Power Syst. **44**(14), 26–34 (2020)
7. Xu, X.L., Song, Y.Q., Yao, L.Z., Yan, Z.: Source-load-storage distributed coordinative optimization of AND (part I): consensus based distributed coordination system modeling. Proc. CSEE **38**(10), 2841–2848+3135 (2018)
8. Xu, X., Song, Y., Yan, Z., Consensus-based source-load-storage optimal dispatch for active distributed network in dynamic multi-agent system. In: 2018 IEEE Power & Energy Society General Meeting (PESGM), Portland, OR, 2018, pp. 1–5 (2018)

Chapter 10
Application of Deep Learning for Registration Between SAR and Optical Images

Wannan Zhang and Yuqian Zhao

Abstract A new matching framework is put forward for registration among optical images along with synthetic aperture radar (SAR). Firstly, picture features are obtained adaptively through convolutional neural networks (CNN). Subsequently, original registration of SAR together with optical pictures is conducted on the grounds of Euclidean distance. In the end, the registration consequence is refined by means of random sample consistent algorithm. The experiment results have illustrated its superior matching execution pertaining to the progressive methods.

10.1 Introduction

In the satellite earth system for observing, both SAR and optical sensors are two dominant pivotal approaches. Two categories of sensors have a great complementation, because they own diverse imaging functions. What's more, the achieved images show diverse features of the observed object. The blend between them has been highly employed in armies of prospects [1, 2]. Additionally, the registration of SAR plus optical pictures is a pivotal basis with a view to integrating. Synchronously, the registration of SAR and optical pictures enhances the precision of SAR picture positioning as well.

SAR possesses the unique benefits, to name only a few, all-weather, multi-frequency band in addition to multi-polarization. It has a certain surface penetration capabilities and detection capabilities for moving targets on the sea and land, but its imaging data has speckle noise and unintuitive interpretation [3]. Currently, the approaches of optical and SAR image registration are chiefly classified into two sorts [4]: (1) The pixel value of the image is straight adopted by the gray-scale way in order to compute the similarity between images. These approaches generally include cross-correlation suiting approaches in addition to least squares. Matching method [5], maximum likelihood method [6], mutual information method [7], divergence

W. Zhang · Y. Zhao (✉)
School of Automation, Central South University, Changsha, China
e-mail: zyq@csu.edu.cn

S. Patnaik et al. (eds.), *3D Imaging—Multidimensional Signal Processing and Deep Learning*, Smart Innovation, Systems and Technologies 348,
https://doi.org/10.1007/978-981-99-1145-5_10

statistics method [8], implicit similarity method, etc. [9]. These methods frequently request that the variance is small among optical and SAR images. Mutual information and other calculation values will not be reflect accurately the degree of correspondence when the resolution difference is large. (2) The feature-based method initially extracts characteristics of the image, to name but a few, points, lines, as well as surfaces. Additionally, it is matched in the spatial position similarity or characteristic depiction similarity [10]. As for feature-based approach, it improves the distinctions from SAR pictures through the enhancement of SIFT and linear features. But it is usually applied for certain specific ground objects such as urban areas, and the promotion ability of such algorithm is poor. There are also some researches concerning multiple registration strategies to improve the effect of registration. It obtains the initial transformation parameters through coarse registration, and then the refined registration is carried out on this basis. Such kind of researches is relatively simple and does not fully complain the respective advantages and the complementary of each other.

A new matching framework for registration among optical pictures as well as SAR is put forward in the article. Firstly, picture characteristics are obtained adaptively through CNN. Then, initial registration of SAR plus optical pictures is conducted on the grounds of Euclidean distance. In the end, the registration consequence emerges from random sample consensus algorithm (RSCA). Experimental consequences prove its ranking matching implementation pertaining to the progressive methods.

10.2 Methodology

10.2.1 Using CNN for Feature Extraction

Currently, CNN has been pervasively applied to image feature extraction together with detection and has obtained exceedingly great consequences [11]. A particular feature of CNN which differs from the conventional algorithms is that it is able to extract characteristics from diverse depth levels of the goal. Diverse convolutional layers in CNN possess diverse feature depiction abilities. In light of the particular demands of remote sensing pictures joint assignments, the way of abstracting layered convolution characteristics is employed in this article, to put it from another angle, the further study CNN way of extracting convolutional layer features of diverse depths is applied. The features of every layer is going to decide the positions of the feature points. Also, the consequences of multiple convolutional layers are integrated with the objective of attaining the position of the characteristic point of the image to be put together.

According to the research of literature [12], there is an advantage in the description of spatial characteristics when it is closer to the previous convolutional layer. In this research, a thoroughly connected layer is not employed. The dominating consideration is the deficiency of its capability of showing the spatial characteristics. Nevertheless, the joint of remote sensing pictures demands to consider not only spatial distinctions but also profound semantic characteristics.

This implies that we ought to think about the spatial position information not to mention the spectral information. A VGG-Net pattern in CNN is adopted, which owns the greatest registration accuracy for location information in the 3rd, 4th as well as 5th convolutional layers. It adopts the yield of the diverse convolutional pool layers near the characteristic points that act as the feature.

By virtue of pooling operation in CNN, spatial resolution is decreased step by step with the growth of depths. The low resolution doesn't conduce to the precision of positioning in the characteristic point space. As a result, the bilinear interpolation way is brought in. Make p stand for the feature layer behind pooling which presents the initial image size after interpolating, the characteristic vector of the ith location is [13]

$$x_i = \sum_k \alpha_{ik} p_k. \tag{10.1}$$

In the formula, α_{ik} means the interpolating weight that relies on the position i and the characteristic vector p_k. The characteristic layer could be acquired.

10.2.2 Improved Euclidean Distance for Matching

As for SIFT picture registration algorithm, the characteristic point matching is a little more complex. Besides, the registration precision and its efficiency are strikingly not high. Of SIFT algorithm, just Euclidean distance is adopted to suit the picture characteristic points. Concurrently, there will exist quite a few error matching point pairs because Euclidean distance merely counts on characteristic points distance similarity. In answer to this issue, a two-layer image feature point matching algorithm is presented, which enhances the Euclidean distance. The algorithm initially employs the distance ratio between the closest and the next closest neighbors as the first level of registration standard.

Let the feature point descriptors in the reference image and the image to be registered be $R_i = (r_{i1}, r_{i2}, \ldots r_{i128})$ and $S_i = (s_{i1}, s_{i2}, \ldots, s_{i128})$, respectively. Similarity measure of any two feature points is [14]

$$d(R_i, S_i) = \sqrt{\sum_{j=1}^{128} \left(r_{ij} - s_{ij}\right)^2}. \tag{10.2}$$

To get the key point descriptor $d\ (R_i, S_i)$ of the pairing, it needs to satisfy

$$\frac{D_\alpha}{D_\beta} < T. \tag{10.3}$$

In them, D_α stands for the S_j closest to R_i in the image with the aim of being registered, D_β stands for the second closest point S_p from R_i in the image with the objective of being registered, and T refers to the rate of closest adjacent Euclidean distance to next closest adjacent Euclidean distance. This value is employed for the sake of assessing the resemblance of two characteristic points which is better than nearest adjacent Euclidean distance. The effect of the local threshold is better. For a reliable match, there must be more nearest neighbor features that are correctly matched than those that are incorrectly matched. The estimation of matching density can also be used as a special case of feature vagueness. By setting Euclidean distance threshold, some wrong matching points can be removed, but some correct matching points may also be eliminated. The threshold is set to 0.7, by which would lead to a reduction of 6%. The right suiting point pairs are also going to decrease 92% of incorrect suiting point pairs.

Secondly, as to the algorithm, the second level of registration dominantly adopts the random sampling algorithm (RANSAC) [15], which conducts the fine registration on the registration consequences of the first layer. In them, the random sampling consistent algorithm is created by Fischler along with a powerful parameter reckon method raised by Kern and Pattichis [16]. Based on this method, the internal constraints of the data set are used to eliminate deviations data point. Specifically, it first focuses on specific questions for designing the objective function when estimating parameters, and then use the iterative estimation method to get the function parameter value of the number. Divide all data points into inner points and outer points. Those points that satisfy the estimation model are the data points, and the others are outer points. The estimated model parameters are continuously entered through the internal points with iterative calculation.

After using the first level of matching, there are still some mismatching points during the process. There exists a kind of algorithm in which cyclic random sampling for transformation is adopted, and this algorithm is just called RANSAC. It can filter out wrong matching data based on the consistency estimation. The advantages of RANSAC algorithm are its reliability, stability, and high accuracy. It possesses powerful tolerance as well as robustness for the sake of extracting noise and feature points inaccurately. Besides, it owns a great capability of deleting mismatches. In consequence, adopting RANSAC so as to deleting the ultimate wrong matching points is going to get the consequence more correctly which highly enhances the registration accuracy. This two-layer feature matching algorithm will be better than using Euclidean distance only without more consumption.

10.3 Experimental Results and Analysis

The relevant test is conducted by us. The test data consists of different characteristics including different resolutions, incidence angles, seasons, etc. The depiction of data aggregation is reflected in Table 10.1. The experiment consequences are reflected not only in Figs. 10.1, 10.2 and 10.3, but also in Table 10.2.

With a view to quantitatively assessing the performances of the registration, the root-mean-square error (RMSE) [17] between the correlative suiting critical points is used, which could be showed as below

$$\mathrm{RMSE} = \sqrt{\frac{1}{n}\sum_{i=1}^{n}\left(x_i - x_i'\right)^2 + \left(y_i - y_i'\right)^2}. \tag{10.4}$$

In the formula, both (x_i, y_i) and (x_i', y_i') mean the coordinates of the ith matching critical point pair; n stands for the entire amount of matching points. What's more, correct matching ratio (CMR) refers to another efficient measure that is stated as

$$\mathrm{CMR} = \frac{\mathrm{correctMatches}}{\mathrm{correspondences}}. \tag{10.5}$$

"Correspondences" refers to the amount of matches through PROSAC; "correctMatches" is the amount of correct matches through deleting the false matches. The consequences of quantitative assessment for every approach are given in Table 10.2.

It can be seen from Table 10.2 that the SIFT algorithm fails to match in heterogeneous image registration, and the correct matching rate obtained by the SIFT-M [18] and PSO-SIFT [19] algorithms is relatively low, and the PSO-SIFT algorithm runs relatively fast. Though some rule of image redundancy removal, the amount of feature point pairs for registration could be dramatically decreased. Prior to assessing affine conversion pattern, of the characteristic point, the initial picture renewal is able to insure registration of multi-phase pictures correctly. Accordingly, the raised algorithm for registration strikingly decreases the run time, which extremely enhances the work ratio of different kinds of image registration.

Table 10.1 Detailed description of dataset

Image no	Image source	Size/(pixel × pixel)	Spatial resolution/m	Date	Location
1	TerraSAR-X	580 × 520	2.5	07/2018	Urban area
	Google Earth	580 × 520	3	05/2017	
2	TerraSAR-X	650 × 500	3	12/2010	River area
	Google Earth	650 × 500	3	09/2012	
3	TerraSAR-X	550 × 460	2	10/2018	Urban area
	Google Earth	550 × 460	3	04/2018	

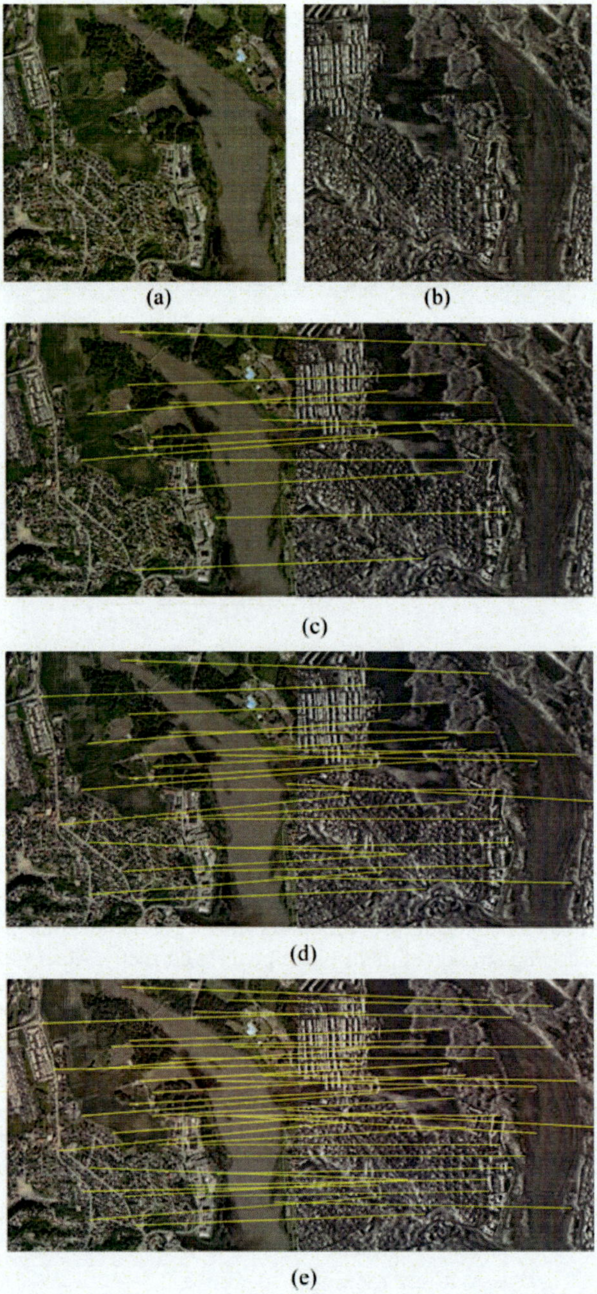

Fig. 10.1 **a** Optical picture; **b** SAR image; matches found in pair 1 using **c** SIFT-M, **d** PSO-SIFT, and **e** the proposed method

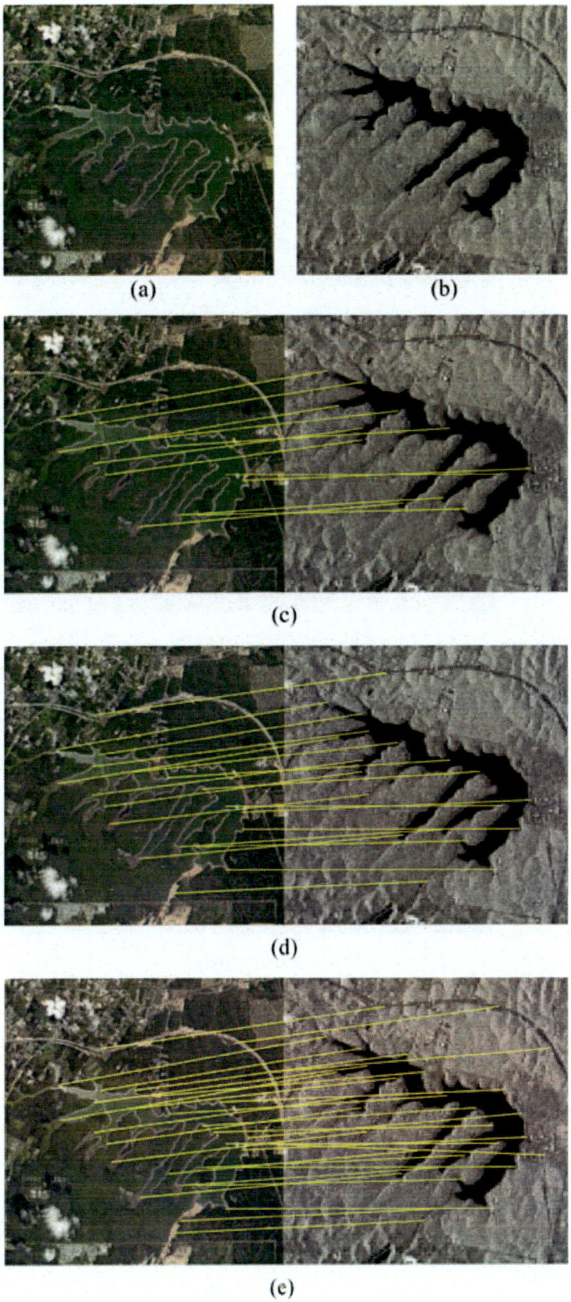

Fig. 10.2 **a** Optical picture; **b** SAR image; matches found in pair 2 using **c** SIFT-M, **d** PSO-SIFT, and **e** the proposed method

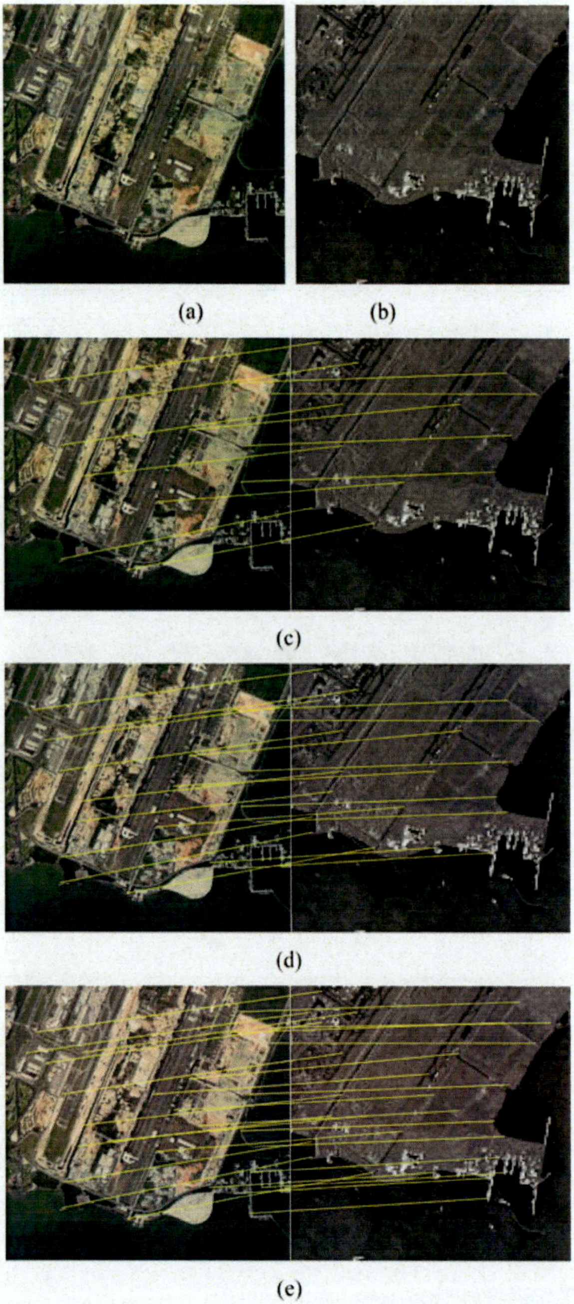

Fig. 10.3 **a** Optical picture; **b** SAR image; matches found in pair 3 using **c** SIFT-M, **d** PSO-SIFT, and **e** the proposed method

Table 10.2 Quantitative comparison of the proposed method with other SIFT-based algorithms

Image no	Method	CMR/%	RMSE/pixel	Running time/s
1	SIFT	–	–	3.36
	SIFT-M	61.14	1.2569	46.75
	PSO-SIFT	68.36	0.9961	42.62
	Proposed	75.45	0.5485	35.82
2	SIFT	–	–	3.77
	SIFT-M	71.28	1.3407	48.91
	PSO-SIFT	69.57	1.4331	41.37
	Proposed	78.09	0.9286	36.06
3	SIFT	–	–	2.43
	SIFT-M	56.81	0.7975	29.63
	PSO-SIFT	48.91	0.9807	23.11
	Proposed	66.56	0.5392	18.39

10.4 Conclusion

One new suiting framework for registration among optical pictures as well as SAR is raised in this research. In the first place, the picture characteristics are gotten by means of CNN. Also, not only initial registration of SAR but also optical images is carried out on the grounds of Euclidean distance. At last, the registration consequence is stated by means of RSCA. The experimental consequences prove the ranking suiting function in respect of progressive methods. The experimental consequences also prove the ranking matching performance pertaining to progressive approaches.

References

1. Zitová, B., Flusser, J.: Image registration methods: a survey. Image Vis. Comput. **21**(11), 977–1000 (2003)
2. Bunting, P., Labrosse, F., Lucas, R.: A multi-resolution area-based technique for automatic multi-modal image registration. Image Vis. Comput. **28**(8), 1203–1219 (2010)
3. Zhu, H., Li, F., Yu, J.M., et al.: Ensemble registration of multi-sensor images by a variational Bayesian approach. IEEE Sens. J. **14**(8), 2698–2705 (2014)
4. Zhang, Q., Cao, Z.G., Hu, Z.W., et al.: Joint image registration and fusion for panchromatic and multi-spectral images. IEEE Geosci. Remote Sens. Lett. **12**(3), 467–471 (2015)
5. Li, J.Y., Hu, Q.W., Ai, M.Y.: Robust feature matching for remote sensing image registration based on Lq-estimator. IEEE Geosci. Remote Sens. Lett. **13**(12), 1989–1993 (2016)
6. Gu, Y.J., Ren, K., Wang, P.C., et al.: Polynomial fitting-based shape matching algorithm for multi-sensors remote sensing images. IEEE Geosci. Remote Sens. Lett. **76**(7), 386–392 (2016)
7. Lowe, D.G.: Distinctive image features from scale-invariant key points. Int. J. Comput. Vis. **60**(2), 91–110 (2004)
8. Bay, H., Ess, A., Tuytelaars, T., et al.: Speeded-up robust features. Comput. Vis. Image Underst. **110**(3), 346–359 (2008)

9. Mikolajczyk, K., Schmid, C.: Scale and affine invariant interest point detectors. Int. J. Comput. Vis. **60**(1), 64–86 (2004)
10. Argenti, F., Lapini, A., Bianchi, T., Alparone, L.: A tutorial on speckle reduction in synthetic aperture radar images. IEEE Geosci. Remote Sens. Mag. **1**(3), 6–35 (2013)
11. Bruzzone, L., Bovolo, F.: A novel framework for the design of change-detection systems for very-high-resolution remote sensing images. Proc. IEEE **101**(3), 609–630 (2013)
12. Mao, X., Shen, C., Yang, Y.B.: Image restoration using very deep convolutional encoder-decoder networks with symmetric skip connections. In: Advances in Neural Information Processing Systems, pp. 2802–2810 (2016).
13. Zhang, K., Zuo, W., Chen, Y., Meng, D., Zhang, L.: Beyond a Gaussian denoiser: residual learning of deep CNN for image denoising. IEEE Trans. Image Process. **26**(7), 3142–3155 (2017)
14. Chandrakanth, R.: Fusion of high resolution satellite SAR and optical images in international workshop on multi-platform/multi-sensor. Remote Sens. Mapp. (2011)
15. Li, W.: A maximum likelihood approach for image registration using control point and intensity. IEEE Trans. Geosci. Remote Sens. **45**(5), 1115–1127 (2004)
16. Kern, J.P., Pattichis, M.S.: Robust multispectral image registration using mutual-information models. IEEE Trans. Geosci. Remote Sens. **24**(18), 3701–3706 (2003)
17. Wang, Z., Bovik, A.C., Sheikh, H.R., Simoncelli, E.P.: Image quality assessment: from error visibility to structural similarity. IEEE Trans. Image Process. **13**(4), 600–612 (2004)
18. Ye, Y., Shan, J., Hao, S., Bruzzone, L., Qin, Y.: A local phase based invariant feature for remote sensing image matching. ISPRS J. Photogramm. Remote Sens. **142**, 205–221 (2018)
19. Selehpour, M., Behrad, A.: Hierarchical approach for synthetic aperture radar and optical image coregistration using local and global geometric relationship of invariant features. J. Appl. Remote Sens. **11**(1), 15–20 (2017)

Chapter 11
Research on Digital Architecture of Power Grid and Dynamic Analysis Technology of Digital Project

Xinping Wu, Gang Wang, Changhui Lv, Bo Zhao, Jianhong Pan, and Aidi Dong

Abstract The traditional project technical and economic evaluation only considers the business needs of the project itself, but does not consider the project construction from the overall situation of power grid construction, which makes it difficult to meet the personalized needs of power grid digital projects. This paper first analyzes the microservice architecture and the characteristics and application scenarios of power grid enterprise middle office. Secondly, aiming at the business demand scenarios such as basic resource operation, power data value-added services, and digital platform ecological construction, the paper studies the design methods of power grid digital business, technology, and security architecture, and designs the mid platform technology and security architecture system of power grid enterprises. Finally, based on the above research results, a dynamic construction and calculation method of digital project evaluation index is proposed to realize the technical and economic dynamic analysis of power grid across business domains.

11.1 Introduction

The traditional project technical and economic evaluation only considers the business needs of the project itself, but does not consider the project construction from the overall perspective of power grid construction [1, 2]. The enterprise middle

X. Wu (✉)
State Grid Economic and Technological Research Institute Co., Ltd, Beijing 102209, China
e-mail: 26261045@qq.com

G. Wang
State Grid Smart Grid Research Institute CO., LTD, Nanjing 210003, China

State Grid Key Laboratory of Information & Network Security, Nanjing 210003, China

C. Lv · B. Zhao
Power Economic Research Institute of Jilin Electric Power Co., Ltd, Changchun 30011, China

J. Pan · A. Dong
State Grid Jilin Electric Power Company Limited, Changchun 130021, China

S. Patnaik et al. (eds.), *3D Imaging—Multidimensional Signal Processing and Deep Learning*, Smart Innovation, Systems and Technologies 348,
https://doi.org/10.1007/978-981-99-1145-5_11

office architecture is a flexible "big, middle, and small front office" organization mechanism, which strengthens the overall service capability of the system through the abstraction of business, data, and technology, and eliminates barriers between various divisions within the system, thus supporting the rapid development of power grid business applications [3, 4]. This paper studies digital business, application, technology, security architecture, and power grid enterprise middle office architecture based on enterprise middle office architecture theory. The dynamic construction and calculation of digital project evaluation indicators are realized based on the grid middle platform architecture.

11.2 Enterprise Middle Office Architecture

(1) Definition and composition of middle platform

The middle office architecture was first proposed by Finland's Supercell Company, which refers to the separation of product technical strength and data operation capability from the front office to become an independent middle office, realizing service sharing, giving the front office greater flexibility and lower change costs, so as to better meet business development and innovation needs. The middle office manages the business platform according to capabilities and services and pursues the integrity of data management and the convenience of business development on the basis of high cohesion and low coupling. The middle office architecture is an organizational mechanism to build a flexible "big middle office and small front office." It strengthens the service ability of the system as a whole through the abstraction of business, data, and technology and eliminates barriers between various divisions within the system. It is a "decentralized" architecture that can adapt to the diversified development of business and provide a clear path for the digital transformation of the organizational structure [5, 6].

(2) Data governance and data middle office

Data governance has always been a very important link in the process of digitalization and informatization. The Data Governance Institute (DGI) proposed the Data Governance Framework in its research report released in 2014, indicating that data governance includes the decision-making power and responsibility system of information related processes, which can help enterprises make better decisions. The data middle platform is to provide full life cycle management of data in the form of services on the basis of data integration, facilitate business construction, and realize the value of data for application business. In recent years, data operations (DataOps) are often mentioned. This concept was included in the technology maturity curve by Gartner in 2018, which is similar to the middle office strategy. DataOps are oriented to data integration and reduce the threshold of data analysis. It is not a specific method or tool, but the integration of various practices and technologies. On the basis of establishing the data analysis model, the data center uses data analysis, data mining, data management, artificial intelligence, and other technologies to establish a unified platform for data sharing and exchange, simplify processes, improve efficiency, and achieve a closed loop between data services and business services [7, 8].

(3) Microservice architecture and business middle platform

The business middle platform is a reusable digital system that supports multiple foreground businesses, and its core is the reuse of digital capabilities in the region. The stable and common business in the system can be separated from the foreground and sunk to the middle platform to improve the efficiency of business application development and iteration speed, so that the front-line business in the foreground can become more agile, quickly adapt to the rapidly changing business needs, and form the microservices architecture. This architecture divides applications into a group of fine-grained services. Compared with the traditional Monolithic Architecture, it has higher maintainability, scalability, and adaptability. Amazon, Netflix, Pay Pal, and other Internet companies first started to try to microservice software systems [9, 10].

11.3 Architecture Design of Power Grid Digital Service

The business reference model of power grid companies is a hierarchical description of the business architecture of power grid companies and a display of the panoramic business view of power grid companies. Assist in diagnosing the health of business operation, optimize business processes, guide the planning, investment and construction of digital projects of power grid companies, avoid excessive digital construction and duplication of functions, help eliminate data redundancy, innovate business models, and provide input and guidance for application system demand analysis and design.

Starting from the business domain, the business functions are refined according to the relevant business capabilities of the business domain through the "top-down" analysis and design method. The business functions are generally composed of multiple business capabilities, which are usually similar to the division of departments within the functional departments in the organizational structure. There is interaction between business domain and business domain. There is cross-department and cross-specialty collaboration. The performance of business functions requires a series of business processes. Each business process is further decomposed into a group of business activities. Finally, a business reference model with four layers of tree structure, namely strategy layer, management layer, design layer, and operation layer, is formed.

With the promotion of the dual carbon goals, the business of power grid companies is also changing. The existing business architecture model needs to be optimized, mainly reflected in the business domain and business functions. Starting from the business strategy of the enterprise, according to the reform of the power grid business, in combination with the reform of electricity sales and measurement, the business domain will be adjusted. At present, the business domains such as human resource management, financial management, material management, operation, and maintenance will be changed with the promotion of the dual carbon goals. According to the work scope of the business department corresponding to each business domain, analyze the business functions, standardize the business functions, adjust the relationship between the business functions and business domains, lay the foundation for the design of enterprise application, data and technical architecture, and guide the digital analysis and design.

11.4 Architecture Design of Power Grid Digitalization Technology

During the construction of the digital system of the power grid company, it is necessary to fully consider the implementation of the application architecture and data architecture according to the content of the business architecture and verify and select the technical framework, database, hardware, network, technical standards, technical components, and other aspects, so as to achieve the goal of business construction and ensure the effectiveness of the information construction investment of the power grid company.

With the promotion of the dual carbon goal, the requirements for business response are getting higher and higher. The existing technical architecture model has been difficult to match the rapidly changing business needs. It is necessary to fully apply the technology such as "Cloud Mobile," optimize the technical architecture reference model, improve the bearing capacity of the integrated platform in all aspects with the mid platform architecture, and achieve the elastic expansion of resources for upper applications to support and drive the power grid development and business innovation goals.

On the whole, the digital grid technology architecture can be divided into six levels: sensing measurement, edge computing, information connection, digital platform, cross-domain intelligence, and system security throughout the whole, as shown in Fig. 11.1.

(1) Sensing and measuring layer

As the basic link of the digitalization of the physical power grid, the sensing and measurement layer is composed of a large number of sensing and measurement devices for electrical and various non electrical links. It is the "nerve endings" of the power grid that transform the physical process into digital signals. Its main tasks include two aspects: first, relying on multi physical quantity sensing devices to obtain more complete basic link information, solving the accuracy problem of models and parameters by establishing the completeness of data, characterizing the complex physical state and dynamic process of the system, providing a basis for accurate state awareness of the power grid, and meeting the requirements of real-time state detection and accurate power prediction of new energy power generation; second,

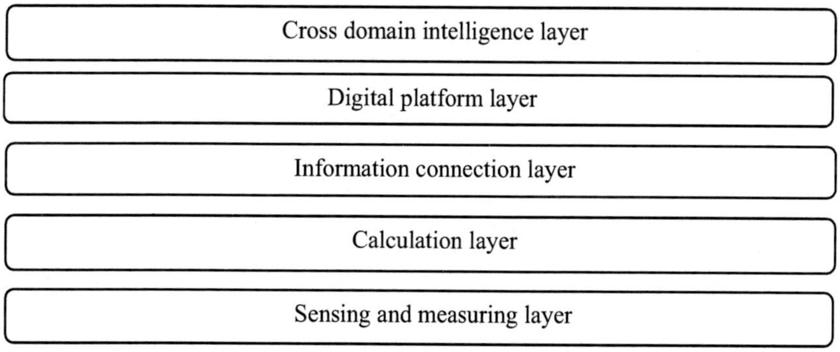

Fig. 11.1 Digital power grid technology architecture diagram

with the help of more low-cost and miniaturized sensing measurement means, it can meet the flexible deployment and application of the grid in multiple application scenarios and complex external environments, support the construction of large-scale and networked sensing measurement system, and realize the comprehensive measurement and perception of the grid to new energy.

(2) Calculation layer

Edge computing provides a targeted solution for the acquisition and utilization of massive data after the digitalization of the physical power grid and constitutes the "low nerve center" of the digital power grid. The edge side of digital power grid is not only the source side of massive data, but also the object side close to a large number of sensing and control targets. The operation data collected by various sensor measurement terminals are collected to the edge computing device through the local communication network, and the rapid analysis and calculation and business services are completed locally; important information obtained from edge computing can be further uploaded to the cloud to complete more complex analysis and decision-making. Relying on edge computing technology, a cloud edge collaborative grid operation architecture will be formed, which can solve the huge resource demand brought by massive data transmission and centralized utilization, endow the grid digital system with more complete local awareness, control and business collaboration optimization capabilities for new energy at the edge, support the safe and efficient consumption of new energy, and provide a more rapid and flexible operation basis for large-scale power systems.

(3) Information connection layer

Information connection is a necessary guarantee to support the two-way flow of power grid data and information between edge computing devices and cloud digital platforms, and constitutes the "neural path" of the power grid digital system. The information connection layer should not only support the diversified needs of digital grid such as access to a large number of edge computing devices, multi-element connection, flexible networking, etc., but also realize efficient transmission of multiple types of data such as pictures, voice, and images; at the same time, it also needs to meet the strict requirements of core businesses such as power grid digital system control and protection on communication real time, reliability, transmission rate, delay jitter, and other indicators, which need to be achieved through the collaborative use of a variety of communication technologies.

(4) Digital platform layer

The digital platform is a powerful software and hardware platform with cloud resource storage, big data processing, data driven analysis, and other capabilities and is the basis for realizing the core functions and intelligence of the digital grid. In terms of hardware, the digital platform will rely on various high-performance computing and data storage and exchange equipment to form a computing resource cluster under the advanced architecture and realize the optimal configuration and coordinated scheduling of hardware resources; in terms of software, we will use big data fusion and mining analysis methods to complete the perception, cognition, and comprehensive presentation of physical power grid based on massive data and establish a digital twin image of power grid operation status and core characteristics. On this basis, the digital platform integrates the grid knowledge obtained from data drive with the original mechanism of the physical grid, providing the grid with forward-looking situation judgment and operation decisions, and ensuring the safe operation of the system after the high proportion of new energy is connected.

(5) Cross-domain intelligence layer

The construction of cross-domain intelligence is the advanced development goal of digital power grid and constitutes the "high-level nerve center" of the digital power grid system. The power grid digital system finds the operation rules and development trend of local systems at the edge, and the digital platform completes the global knowledge discovery at the system level in the cloud. Based on this, the cross-domain intelligence layer will first face different application needs and form business intelligence through learning specific data; Then, the knowledge of different business domains is correlated with each other by means of data networks to form the knowledge network and knowledge atlas of the power grid; finally, realize the migration of knowledge among fields, master the interaction and synergy between various business applications of the power grid, and achieve intelligent collaboration across business domains.

11.5 Midrange Architecture of Power Grid Enterprises

The power grid company grasps the direction of digital development, conforms to the trend of digital transformation and middle platform development, speeds up the construction of enterprise middle platform, focuses on building enterprise middle platform, enables power grid production, enables customer service, enables enterprise operation, and helps to create an energy Internet ecology.

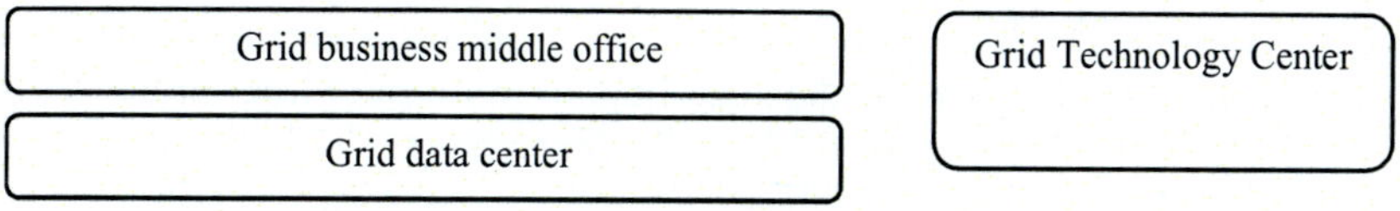

Fig. 11.2 Middle office architecture of power grid enterprises

Based on the grid cloud, the middle office of power grid enterprises is an enterprise level capability sharing platform, which enables cross-business reuse and global data sharing in business, supports rapid and flexible construction of front-end applications, supports rapid business development, agile iteration, and on-demand adjustment, enables business development and management innovation of the company, achieves comprehensive precipitation of application services and technical capabilities in major business areas, and helps promote the high-quality development of the company and power grid. The middle office of power grid enterprises is shown in Fig. 11.2.

(1) Grid business middle office

The business middle platform is an enterprise level business capability sharing platform, which precipitates the core business processing capacity of the enterprise into each sharing capability center, and continuously improves the efficiency of business innovation. Integrate the resources and capabilities that need to be reused across disciplines in the core business of the power grid company, provide "agile, fast, low-cost" innovation capabilities and unified enterprise level shared services, eliminate business breakpoints, and avoid repetitive maintenance, mainly focusing on supporting front-end business processing.

The power grid business middle office is composed of customers, power grid resources, finance, and other middle offices, forming an enterprise level sharing capability center to provide unified business service scheduling for the front office and will continue to expand, iterate, and improve in the later period.

(2) Grid data center

The power grid data middle office is an enterprise level data capability sharing platform, which gathers and integrates the data of various disciplines and units of the power grid into enterprise level data services. Through hierarchical and horizontal decomposition, the data is aggregated, stored, integrated, analyzed, and processed to precipitate public data capabilities. Service encapsulation is carried out according to business scenarios to form enterprise level data services, support the agile iteration and rapid construction of front-end applications, and realize data value sharing.

(3) Grid technology center

The power grid technology platform is an enterprise level technology capability reuse platform, which provides "unified architecture, advanced technology, and intelligent service" enterprise level basic public service capabilities for power grid businesses. The technical middle office includes video, geographic services, and other technical platforms. Through the continuous platform precipitation of technical capabilities, the technical capabilities of the power grid are separated from the business capabilities, and the precipitation forms the service capabilities of equipment access services, digital identity opening, unified map services, and so on. The technology innovation and sharing services of "unified, easy to use, and robust" are provided for the business middle office, data middle office, and front office in a product-based manner, Assist the rapid construction of enterprise digital applications and achieve strong support.

(4) Relationship between power grids

The data middle office receives the data transmitted by the middle office of the business, cleans, processes, and analyzes the data through big data algorithm, modeling and other technologies, and the results support the middle office of the business for the front-end business application to call immediately, while the new data generated by the front-end business application flows to the data middle office to form a closed loop; the technical middle office provides unified technical services for the business middle office and data middle office.

11.6 Dynamic Construction and Calculation of Digital Project Evaluation Index Based on Grid Middle Platform Architecture

The dynamic construction and calculation of indicators are shown in Fig. 11.3.

First of all, according to the midrange architecture of power grid enterprises, the design principles of business, application, data, technology, and security architecture are sorted out, and the Delphi method is used to form a quantitative measurement index system of digital development demand;

Secondly, the expert method and fuzzy information axiom method are used to calculate the quantitative measurement index value of digital development demand in combination with the business architecture, application architecture, and security architecture of the middle office architecture, focusing on measuring the cross-business domain situation of business indicators and dynamically obtain the empowerment results according to the architecture analysis;

Then, based on the grid business middle office, data middle office and technology middle office, the association relationship and coefficient table between the quantitative measurement indicators, and the evaluation objectives are established, and the

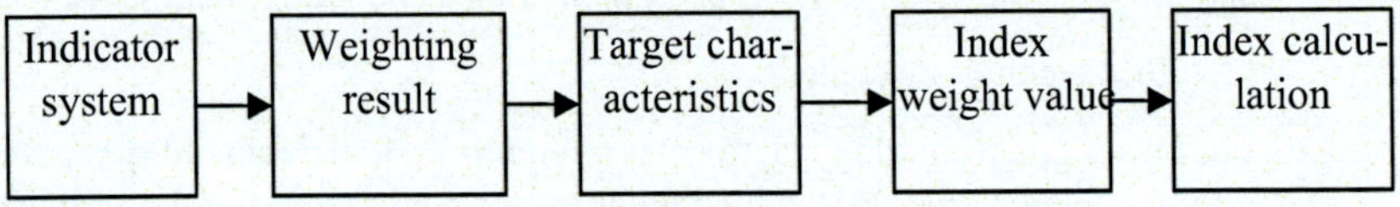

Fig. 11.3 Dynamic construction and calculation flowchart

business needs are transmitted to the characteristics of the project objectives to be evaluated in combination with the business architecture;

Then, obtain the compliance correspondence of the functional architecture, technical architecture, and security architecture of the digital project to be evaluated from the analysis of the indicator construction link, call the evaluation indicator weight value, and multiply and add the weight according to the calculation results of the correlation coefficient;

Finally, combine and normalize the evaluation target expert weighting results with the index weight calculation results to obtain the heaviest evaluation index calculation value, realize the dynamic construction and calculation of indicators based on the power grid mid platform architecture, and support project planning, design, and optimization sorting.

11.7 Conclusions

This paper first analyzes the definition and composition of enterprise middle office, data governance and data middle office, microservice architecture and business middle office characteristics, and application scenarios. Secondly, in view of the business demand scenarios such as basic resource operation, power data value-added services, and digital platform ecological construction, the research focuses on the architecture design, architecture design, technical architecture design, and security architecture design of power grid digital business, proposes architecture design methods, proposes sensor measurement, edge computing, information connectivity, digital platforms, and cross-domain intelligent technology architecture systems, and proposes terminal security access, information physical security protection, data security and protection, security integration security architecture. Then, based on the research of architecture technology, the mid-platform architecture of power grid enterprises is studied. Finally, based on the grid mid platform architecture, a dynamic construction and calculation method of digital project evaluation index is proposed to realize the dynamic analysis of grid technology and economy across business domains. Compared with the traditional method, the method proposed in this paper considers the economic and technical analysis of a single project from the overall situation of power grid digital construction, avoiding the local optimization of the traditional method and the unreasonable situation of the overall consideration.

Acknowledgements This work is supported by Science and Technology Project of State Grid Corporation of China (Research on the evaluation system and technology tools of digital technology and economy of enterprises, No. 5700-202129180A-0-0-00).

References

1. Xiang, M., Chenhan, S.: Practice of enterprise architecture method based on TOGAF. Digit. Commun. World **4**, 194–196 (2017)
2. Yoo, Y., Henfridsson, O., Lyytinen, K.: Research commentary—the new organizing logic of digital innovation: an agenda for information systems research. Inf. Syst. Res. **21**(4), 724–735 (2010)
3. Paulus, R.D., Schatton, H., Bauernhansl, T.: Ecosystems, strategy and business models in the age of digitization-how the manufacturing industry is going to change its logic. Procedia CIRP **57**, 8–13 (2016)
4. Yongwei, L., Qian, S., Bo, L.: Enterprise information architecture design. China Sci. Technol. Inf. **18**, 43–44 (2015)
5. Li, Z.: Research and application of digital transformation technology in power enterprises. China New Commun. **01**, 127–128 (2020)
6. Bin, C., Xiaoyi, Y.: Research and practice of power enterprise information management based on enterprise architecture. Enterp. Manag. **S2**, 526–527 (2016)
7. Degree in the manufacturing industry in Germany. Procedia CIRP **57**, 14–19 (2016)
8. Tong, Y., Li, Y.: The Evaluation of enterprise informatization performance based on AHP/GHE/DEA. In: 2007 International Conference on Management Science and Engineering, pp. 149–155. IEEE (2007)
9. Jun, W., Yongqiang, H.: Discussion on information architecture management of power enterprises. Power Inf. Commun. Technol. **14**(6), 14–17 (2016)
10. Jiachen, T.: Research on enterprise information architecture design of the company. Wuhan University of technology (2016)

Chapter 12
Research on Characteristics and Architecture Application Technology of Power Grid Digital System

Xinping Wu, Gang Wang, Changhui Lv, Bo Zhao, Jianhong Pan, and Aidi Dong

Abstract Digital technology plays an increasingly important role in the development of the company and the power grid. The traditional project economic and technical analysis method mainly uses fixed indicators to carry out technical and economic analysis, which cannot achieve dynamic indicator adjustment, and the analysis quality is poor. Traditional methods cannot meet the technical and economic evaluation needs of cross-business digitalization in digital construction. This paper studies the enterprise architecture theory, the connotation and technical characteristics of the power grid digital system, and summarizes the characteristics of the power grid digital system as mass data, extensive connection, software definition, data drive, and cross-domain intelligence. Based on the technical characteristics of the new architecture, this paper proposes a technical scheme of enterprise level digital architecture for power grid business. According to the digital needs and construction status of enterprises, combined with the characteristics of the digital report to be reviewed and its business scenarios, the technical and economic dynamic analysis method of digital projects based on the grid architecture can realize dynamic intelligent evaluation oriented to the grid digital scenarios, improve the technical and economic evaluation efficiency of the grid enterprises' digital benefits, and improve the collaborative management and control level of the grid digital system construction.

X. Wu (✉)
State Grid Economic and Technological Research Institute Co., Ltd, Beijing 102209, China
e-mail: 26261045@qq.com

G. Wang
State Grid Smart Grid Research Institute CO., LTD, Nanjing 210003, China

State Grid Key Laboratory of Information & Network Security, Nanjing 210003, China

C. Lv · B. Zhao
Power Economic Research Institute of Jilin Electric Power Co., Ltd, Changchun 30011, China

J. Pan · A. Dong
State Grid Jilin Electric Power Company Limited, Changchun 130021, China

S. Patnaik et al. (eds.), *3D Imaging—Multidimensional Signal Processing and Deep Learning*, Smart Innovation, Systems and Technologies 348,
https://doi.org/10.1007/978-981-99-1145-5_12

12.1 Introduction

The informatization construction of the power grid company has made great progress, gradually realizing the transformation of the main business from decentralization to centralization, offline to online, and island to integration, better supporting various production and operation management activities, and laying a good foundation for the digital development of the power grid company [1, 2]. However, with the gradual improvement of grid intelligence, digital technology plays an increasingly important role in the development of the company and the grid. However, the traditional economic and technical analysis methods mainly use manual methods to carry out fixed index evaluation, which cannot achieve automatic intelligent evaluation. In the face of the technical and economic evaluation needs of multi-scene complex grid applications, multi-dimensional technical and economic comprehensive evaluation, intelligent identification of digital technology target characteristics, cross-business digital needs, and other technical and economic evaluation needs of digital construction, traditional evaluation methods have been unable to meet the needs of cross business digital construction evaluation [3, 4]. It is urgent to study the digital architecture technology of power grid to solve the cross service analysis problem of power grid.

12.2 Enterprise Architecture Theory

Enterprise architecture is often called information architecture. It is the top-level design of enterprise information, the best practice to guide enterprise information construction, and a set of relatively perfect theories and methods from planning, design to implementation. Literally, the enterprise architecture is composed of 'enterprise' and 'architecture' [5, 6]. In the Modern Chinese Dictionary, the enterprise (Enterprise) is interpreted as a department engaged in production, transportation, trade, and other economic activities, such as factories, mines, railways, and companies [7, 8]. Generally speaking, 'enterprise' refers to a commercial organization that is composed of a set of identifiable and interacting business capabilities and operates according to certain organizational rules, while architecture can be traced back to the building field at the earliest. Like architecture, it is composed of bricks, concrete, and other basic elements to carry the pressure transmitted by the upper layer. In Weber's dictionary, the definition of architecture is 'a form or framework that is the result of a conscious process; a unified or organized form or structure; and the art or science of architecture'. In this definition, it is clear that the key part of architecture is a conscious and organized method that has a specific structure, reflects some daring things, and aims at things [9, 10].

12.3 Research on Characteristics of Power Grid Digital System

Digitalization is to simulate and virtualize the physical system in the computer system, reflect the physical world in the computer system, use digital technology to drive organizational business model innovation, drive the reconstruction of the business ecosystem, and drive the great change of enterprise services. Digital transformation is a systematic transformation process that takes data as the core element, takes value creation as the purpose, integrates and applies digital technology, and reconstructs management mode, business mode, business mode, governance form, and cultural concept. The construction of power grid digitalization refers to various digitalized construction work to improve the company's grass-roots first-line service ability, core business enabling ability, enterprise transformation driving ability, and industrial upgrading leading ability in the form of power grid digitalization projects through informatization, networking, and intelligence.

The construction of power grid digitalization system covers the digitalization of power grid in physical space and the intellectualization of power grid in digital space. It is a highly complex systematic project. Among them, the digitalization of the physical power grid is the necessary basis for the construction of the digital power grid, and the digital platform is an important way to build the grid intelligence in the digital space.

The digitalization of the physical power grid means the common transformation of the primary and secondary systems, which is mainly reflected in three aspects: The first is the digitalization of the power grid state, especially after the application of a large number of new energy power generation, power electronic devices, new electric equipment, etc., and the power grid state will be electrical oriented, and optical, thermal, fluid, gas, mechanical and other multiple physical processes coexist, requiring high-frequency, multi-modal data acquisition to achieve the power grid state perception of equipment status, transaction status and management status to comprehensively improve the observability of the power grid; the second is the digitalization of the power grid energy flow, that is, through the extensive deployment of flexible and controllable equipment, the energy flow and control response characteristics of the physical power grid can be flexibly defined by the digital space to improve the controllability of the digital power grid, which will bring subversive changes to the operation mode of the traditional power grid; the third is the digitalization of the power network, that is, the construction of an information network corresponding to a complex physical power grid, the formation of a wide range of information connections between the massive elements of the power grid, the enhancement of the power grid discovery, early warning risk and active protection capabilities, and the improvement of the reliability and resilience of the power system.

The power grid digital system aims to build a holographic mapping of complex physical entities from real space to virtual digital space, depict, and simulate the real-time state and dynamic characteristics of the physical power grid in the digital space, so as to complete various analysis and research that are difficult to carry out in the real world in the virtual environment and support various business applications. Its main features include (1) accurate mirror image, that is, digital presentation of the real-time status of the physical grid; (2) observable panorama, providing full cycle and deep level system state observation; (3) dynamic evolution shows the dynamic process of system characteristic evolution in digital space through the feedback of operation information; and (4) virtual real interaction, relying on a two-way information communication network to achieve data aggregation and issue decision instructions.

The power grid digitalization system is the overall appearance of the physical power grid after digitalization and is also the key to forming and exerting the value of the power grid digitalization system. It can achieve data convergence, fusion, presentation and interaction in the digital space, thereby supporting the refinement of the mechanism, rules and knowledge of the power grid based on data, building power grid intelligence driven by data, and completing the system organization and functional operation of the power grid as a whole with data as the core. Among them, the core role of data comes from two aspects: first, due to the small time constant, wide frequency distribution, hybrid dynamic processes and difficulty in obtaining accurate models of the power grid in the new form, relevant technologies and research methods are required to develop from 'model equivalence' to more 'data driven'; second, due to the significant increase in the volatility of the power grid in the new form, the role of source network load storage is integrated and transformed, and energy and information are interwoven and interacted. When solving the giant system problems of multiple complex objects in different space–time scales and different movement processes, it is more necessary to rely on global data as the research basis and link.

Massive data and extensive connectivity are the most basic characteristics of the digital grid at the physical level, which forms the basis of the digital grid's capabilities of 'knowledge discovery' and 'knowledge migration' in the digital space, enabling the digital grid to complete the whole process from physical processes to 'signals,' 'data,' and 'information,' and further refining them into business 'knowledge' and 'intelligence,' ultimately forming the 'wisdom' of the digital grid. The characteristics of the power grid digital system can be summarized in the following five aspects, as shown in Fig. 12.1.

(1) Massive data. The digital power grid does not aim at the specific 'business domain' of a single service traditional power grid, but coordinates the needs of different business domains with public information sources, data resources, and data processing capabilities. This first requires that the digital grid can obtain as rich, comprehensive, relevant, and systematic information measurement as possible in the physical world, not limited to the traditional few measurement

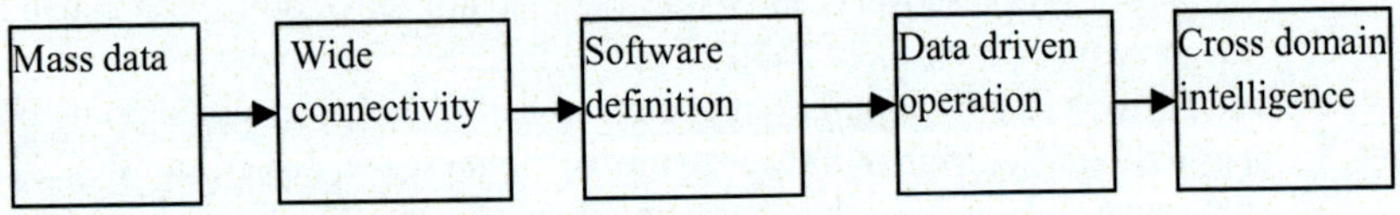

Fig. 12.1 Characteristic diagram of power grid digital system

nodes and separate electrical measurement. Therefore, the data volume of the digital grid will grow explosively in many dimensions, such as spatial breadth, measurement frequency, and data type.

(2) Wide connectivity. The primary equipment and grid of traditional power grid realize the electrical connection between source and load; on this basis, the digital power grid will further rely on advanced information and communication technologies to build an information network corresponding to the physical power grid, forming a two-way information connection with extensive horizontal coverage and full vertical connectivity, so that the physical elements in the power grid not only influence each other electrically, but also can be interconnected and coordinated at the information level. The extensive connection endows the digital power grid with the essential characteristics of large-scale information physical system, making it possible to uniformly perceive and control from the system level.

(3) Software definition. The information physical coupling characteristics of the digital power grid system enhance the influence of the secondary system on the primary system. At the same time, various advanced power electronic devices and controllable equipment provide the digital power grid with the basis for the implementation of complex operation control strategies. This enables the digital grid to meet the needs of multiple types of business with universal computing power in the information link, and to achieve flexible configuration and rapid deployment of policies for different scenarios through software definition. In addition, the software definition also gives the power grid the ability to flexibly define the operating state of the physical system in the digital space, and realizes the closed loop of digital perception physics and digital definition physics.

(4) Data driven. Massive data and extensive connections have greatly improved the data resources obtained by the digital power grid, enabling it to break the boundaries of the traditional grid business domain from the data level and achieve the fusion and refinement of multi-source data in the digital space, so as to more accurately and perfectly depict the operation state and behavior laws of the power grid. On this basis, we will use public data resources and powerful computing power to serve the business needs of the digital grid, achieve 'data driven' grid operation and control, maximize the use of 'data' value, and drive the digital grid to form a high-level business 'intelligence.'

(5) Cross-domain intelligence. The informatization and digitalization development of the traditional power grid takes business as the main line and serves specific business needs, forming an information automation system facing different business domains. In addition to building specific business level intelligence, the digital grid is more important to complete the fusion mining of cross-business data, the construction of knowledge discovery and knowledge maps, and realize the cross-domain migration of 'knowledge' and 'intelligence' between different businesses and fields. This enables the grid digital system to meet the requirements of business intelligence, and at the same time, it also has the ability to actively coordinate different businesses at the system level, so that grid

knowledge can be migrated and reused in different business areas, thus greatly improving the overall 'smart' level of the grid.

12.4 Digital Architecture Design of Power Grid

A metamodel is a model that describes a model. In the field of enterprise architecture, it is a unified definition of various concepts appearing in the architecture. It will sort out the objects related to enterprise digitalization (such as business processes, functions, data entities, and nodes) and define the relationships between these object elements, so as to facilitate mutual reference and verification between objects. The architecture metamodel supports enterprise architecture related personnel to communicate and use the same concepts and dictionaries, so that the architecture information can be saved in a structured form.

Based on the hierarchical application and technical characteristics of the new architecture in the digital construction, a technical scheme for the construction of enterprise level digital architecture for power grid business is proposed, as shown in Fig. 12.2.

First, because the existing information architecture elements are difficult to meet the digital needs, based on the massive data, extensive connectivity, software definition, data drive, cross-domain intelligence characteristics of the power system, as well as the design contents of the digital architecture element model, business architecture, application architecture, data architecture, and technical architecture, the multi-agent modeling method is used to carry out the digital architecture model simulation for the characteristics of the power grid digital system.

Second, based on the characteristics of the power system and the mid platform architecture, an enterprise level capability sharing service adapted to the digital requirements is built from the technical level to achieve cross-business reuse of digital capabilities and global data sharing. Combined with the business requirements of new power system, a common service precipitation mechanism oriented to the middle platform of power system is established. Among them, the principle of power system enterprise level reuse refers to providing power system service reusability through sharing services, realizing resource reuse and application reuse, meeting the requirements of business or data sharing reuse, and realizing service invocation of all lines and businesses of the company.

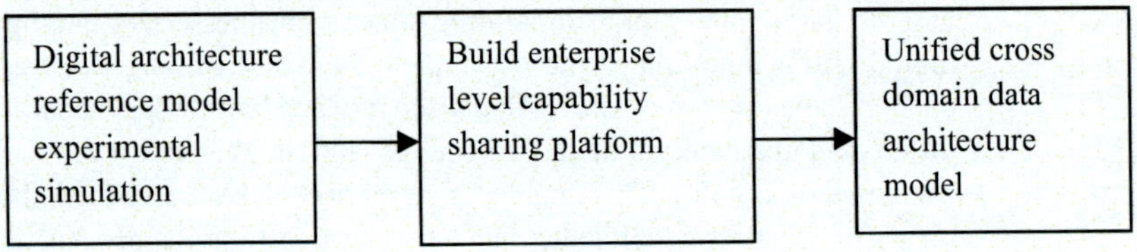

Fig. 12.2 Technical scheme of enterprise level digital architecture construction of power grid

Third, from the data level, a unified cross-domain data architecture model is proposed to adapt to the cross-domain data reuse of the power system and improve the data value sharing capability of the data center. Based on the digital business elements of the power system, a unified cross-domain data model design study is carried out to achieve the unified storage, management, and service under the dual carbon goal, so as to establish and improve the cross-domain digital resource management system, sharing mechanism and methods. Based on power system business characteristics and multi-agent model building method, a unified cross-domain data middle platform is built to form a power system cross-domain data model.

12.5 Technical and Economic Dynamic Analysis of Digital Projects Based on Power Grid Architecture

The key of enterprise digital project review is to adjust the impact of development demand differentiation under different life cycles and business scenarios on the weight of evaluation indicators. The analysis method is shown in Fig. 12.3.

The first step is to dynamically generate the digital project evaluation index system based on the enterprise's digital needs and construction status, combined with the characteristics of the digital report to be reviewed and its business scenarios, using the grid digital architecture technology and multi-agent to analyze the characteristics of the business architecture, analyze the contribution of the enterprise's mid platform and cross-domain data, and combine the importance of the business;

Second, based on the characteristics of the digital system of power grid projects, the key indicators in the evaluation system are selected by using the gray correlation analysis method and expert judgment method;

The third step is to score the quantitative indicators of the digital development needs of enterprises by experts and process the data through the fuzzy information theory to obtain the demand pressure index;

The fourth step is to obtain the correlation analysis result between the demand pressure index and the characteristics of the evaluation target by using the order relation method, call the weight value of the evaluation index of typical scenarios, and modify the weight value of the digital project target characteristics, so as to modify the weight value of the evaluation index;

The fifth step is to conduct technical and economic analysis of the project according to the evaluation index and weight, and determine the project ranking.

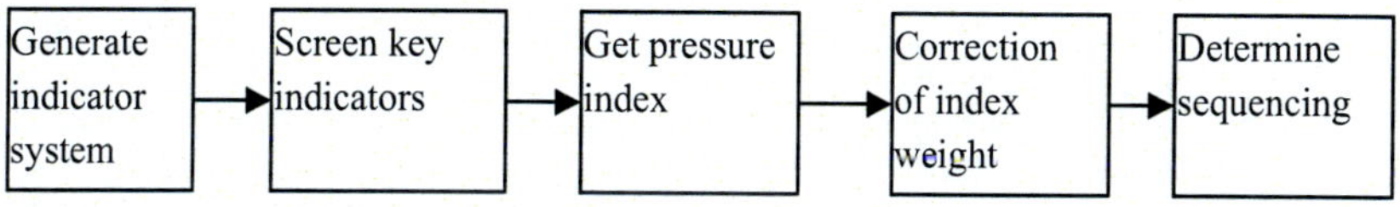

Fig. 12.3 Analysis steps based on power grid architecture

12.6 Conclusion

This paper studies the enterprise architecture theory, the connotation, and technical characteristics of the power grid digital system and summarizes the characteristics of the power grid digital system as mass data, extensive connection, software definition, data drive, and cross-domain intelligence. Based on the technical characteristics of the new architecture, an enterprise level digital architecture technology scheme for power grid business is proposed. According to the digital needs and construction status of enterprises, combined with the characteristics of the digital report to be reviewed and its business scenarios, the technical and economic dynamic analysis method of digital projects based on the grid architecture can realize dynamic intelligent evaluation for the grid digital scenarios, improve the quality of the digital technical and economic evaluation of power grid enterprises, and improve the collaborative construction level of the grid digital system construction.

Acknowledgements This work is supported by Science and Technology Project of State Grid Corporation of China (Research on the evaluation system and technology tools of digital technology and economy of enterprises, No. 5700-202129180A-0-0-00).

References

1. Jun, W., Yongqiang, H.: Discussion on information architecture management of power enterprises. Power Inf. Commun. Technol. **14**(6), 14–17 (2016)
2. Paulus, R.D., Schatton, H., Bauernhansl, T.: Ecosystems, strategy and business models in the age of digitization-how the manufacturing industry is going to change its logic. Procedia CIRP **57**, 8–13 (2016)
3. Yongwei, L., Qian, S., Bo, L.: Enterprise information architecture design. China Sci. Technol. Inf. **18**, 43–44 (2015)
4. Li, Z.: Research and application of digital transformation technology in power enterprises. China New Commun. **01**, 127–128 (2020)
5. Bin, C., Xiaoyi, Y.: Research and practice of power enterprise information management based on enterprise architecture. Enterp. Manag. **S2**, 526–527 (2016)
6. Xiang, M., Chenhan, S.: Practice of enterprise architecture method based on TOGAF. Digit. Commun. World **4**, 194–196 (2017)
7. Yoo, Y., Henfridsson, O., Lyytinen, K.: Research commentary—the new organizing logic of digital innovation: an agenda for information systems research. Inf. Syst. Res. **21**(4), 724–735 (2010)
8. Bogner, E., Voelklein, T., Schroedel, O.: Study based analysis on the current digitalization degree in the manufacturing industry in Germany. Procedia CIRP **57**, 14–19 (2016)
9. Tong, Y., Li, Y.: The evaluation of enterprise informatization performance based on AHP/GHE/DEA. In: 2007 International Conference on Management Science and Engineering, pp. 149–155. IEEE (2007)
10. Jiachen, T.: Research on enterprise information architecture design of the company. Wuhan University of technology (2016)

Chapter 13
Investigation of Vessel Segmentation by U-Net Based on Numerous Datasets

Zhe Fang, Hao Jiang, and Chao Zhang

Abstract The segmentation results of retinal vessels can be an essential foundation for diagnosing and detecting numerous clinical medical symptoms. The work in this paper is to first train on the public dataset DRIVE to obtain the U-Net model. Secondly, the model is applied to private databases to segment fundus vascular images. Finally, deep learning metrics are evaluated on the U-Net model and test results. The experiment results demonstrate the effectiveness of U-Net on vessel segmentation tasks.

13.1 Introduction

Research shows that various eye diseases can cause retinal blood vessels to develop deformities, hemorrhages, and other problems. However, analysis of fundus images can diagnose a variety of ocular and even physical diseases [1]. Experts and physicians can accurately detect many retinal diseases monitoring alterations in retinal vascular patterns, such as branching patterns, width, shape, and curvature. In the conditions above, when the medical standard was not well-developed, the determination of such diseases can only be made by visual observation of medical experts. But the images of the retinal vessels in the fundus could be more transparent with this method, resulting in a considerably reduced contrast between the target vessels and the background, mainly when medical imaging is subject to external interference, which analyzes retinal fundus vessels even harder to perform [2].

Contemporary imaging techniques have been capable of obtaining the vascular network structure of the retina. Deep learning algorithms have been rapidly developed and implemented for fundus image segmentation in the past decade. Numerous

Z. Fang · H. Jiang
Yunnan Normal University, Kunming 650000, Yunnan, China

C. Zhang (✉)
Yunnan Key Laboratory of Optoelectronic Information Technology, Yunnan Normal University, Kunming 650000, Yunnan, China
e-mail: zhchynnu@foxmail.com

S. Patnaik et al. (eds.), *3D Imaging—Multidimensional Signal Processing and Deep Learning*, Smart Innovation, Systems and Technologies 348,
https://doi.org/10.1007/978-981-99-1145-5_13

publications have been published on deep learning applications in machine vision [3] and medical image analysis [4].

The work to be accomplished in this paper is inspired by numerous previous studies and decreased much work by using the classical image segmentation algorithm U-Net. Fundus images have the advantages of being low cost, non-invasive, easy to obtain, and reusable, which has aroused a trend in scientific research regarding retinal vascular segmentation.

Although numerous researchers have identified various unsupervised learning-based algorithms to segment retinal vascular images, there is space to improve the segmentation accuracy relative to micro-vessels and low-contrast images. Literature [5] proposed a vessel segmentation method with the U-Net architecture, which is the foundation for the experiments conducted in this paper. Literature [6] integrated the U-Net and multi-scale filtered vessel algorithms. It used a U-shaped network structure to bridge the output of the encoding layer with the decoding layer, solving the low-grade information-sharing problem more effectively [7]. Literature [8] combined the U-Net network and dense network for retinal fundus vessel segmentation and got superior accuracy, sensitivity, and specificity [7]. Many scholars have favored the U-Net network architecture.

In this study, the U-Net network architecture trained on the DRIVE [9], and private datasets had superior segmentation performance and accuracy. It provides a deeper understanding of the challenges of retinal fundus vascular segmentation in this learning process. Simultaneously, this could inspire future scientists to invest more in this work.

We acquaint our assignment as follows. Section 13.1 is an introduction. This section focuses on the implications of retinal vessel segmentation and related research efforts, followed by a brief description of our experiments. Section 13.2 is an introduction to the U-Net model of deep learning. Section 13.3 provides details of our implementation of U-Net model building and training based on the DRIVE databases. Section 13.4 presents the results of our predictive generation of fundus vessel segmentation on the private databases. Section 13.5 draws a conclusion of our work in this paper.

13.2 Introduction to Deep Learning U-Net Model

The U-Net network structure includes a systolic path for capturing the context and an extended symmetric approach for precise positioning [10], resulting in a U-shaped structure. The architecture of the U-Net can be seen in Fig. 13.1.

The deep learning U-Net model is often used in the research on biomedical imaging. As far as we know, the model of the U-Net network is becoming one of the most commonly used and simplest segmentation models owing to its simplicity, efficiency, understandability, and ease of construction. In addition, it can tackle lacking images in biomedicine and achieve a low error rate.

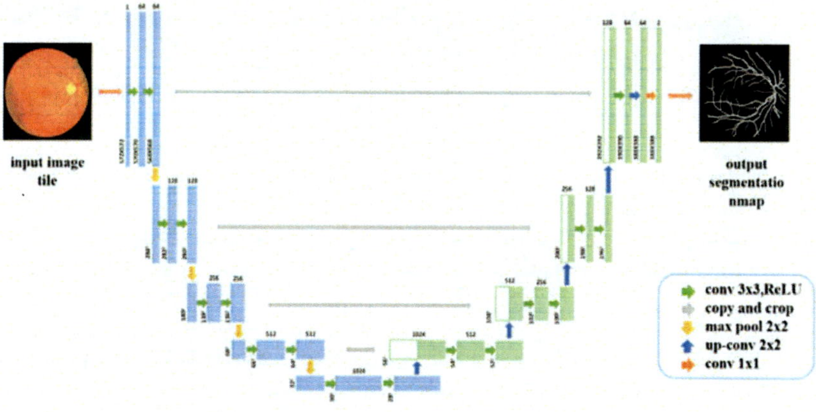

Fig. 13.1 U-Net architecture

In this model, contraction path is deployed to acquire information about the context, whereas expansion path enables precise positioning. Where the loss function is cross-entropy, stochastic gradient descent is used for optimization, and the activation function after each convolutional layer is the rectifier linear unit. The dropout of 0.2 is used between two consecutive convolutional layers whereby the convolutional layers can get a more precise output based on this information.

13.3 Construction and Training of U-Net Model

13.3.1 Datasets

The DRIVE database includes 40 images, with half of the training images and half of the test images [7]. A manual segmentation image of the retinal vessels and an eye contour image are provided for each image; refer to Fig. 13.2.

13.3.2 Data Processing

To perform the data processing, we based on the preprocessing method proposed in the literature [8] and performed the following steps:

(a) Gray-scale transformation: Due to the insufficient brightness range of the image or non-linearity will make the contrast of the image unsatisfactory. Adopt the image gray value transformation method, which means to change the dynamic range of image gray scale by changing the gray value of image pixels, and thus enhance the contrast of the image. Assuming [11] that the original image (pixel

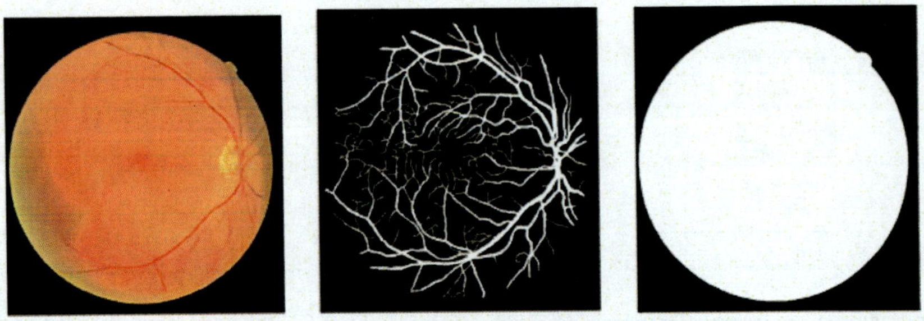

Fig. 13.2 **a** Fundus images of training set from DRIVE databases. **b** Manually annotated vascular images. **c** Eye contour image

gray value) is $f(m, n)$ and the processed image (pixel gray value) is $g(m, n)$, the contrast enhancement can be expressed as:

$$g(m, n) = T[f(m, n)] \tag{13.1}$$

where T (.) denotes the gray-scale transform function of the enhanced image and the original image.

(b) Standardization: Converts the eigenvalues of the data to the same magnitude without changing the ordering of the values in the original data [12]. Standardized data can speed up the convergence of the model and improve the accuracy of the model as well.

(c) Contrast-limited adaptive histogram equalization: Dividing the image into sub-blocks to improve the local contrast of the image, thus performing histogram equalization on the sub-blocks [13]. Then, crop the histogram got from the statistics in the sub-block to keep its magnitude below a certain upper limit. Eventually, distribute this cropped value evenly over the entire gray-scale interval to ensure that the total histogram area remains constant, thus overcoming the problem of over-amplified noise.

(d) Gamma transform: Through nonlinear transformation, it adjusts the contrast of overexposed or underexposed (too dark) gray-scale maps, resulting in enhanced overall image detail. The mathematical model of this principle is as follows [14]:

$$S = Cr^{\gamma} \tag{13.2}$$

where r is the input value of the gray-scale image (the original gray-scale value) and takes the value range [0, 1]. S is the gray-scale output value after gamma transformation. C is the gray-scale scaling factor, usually taken as 1. The γ is the gamma factor size.

Then, we used random slicing to amplify the data from the DRIVE dataset consisting of a mere 20 fundus images and extracted 9,500 random patches from

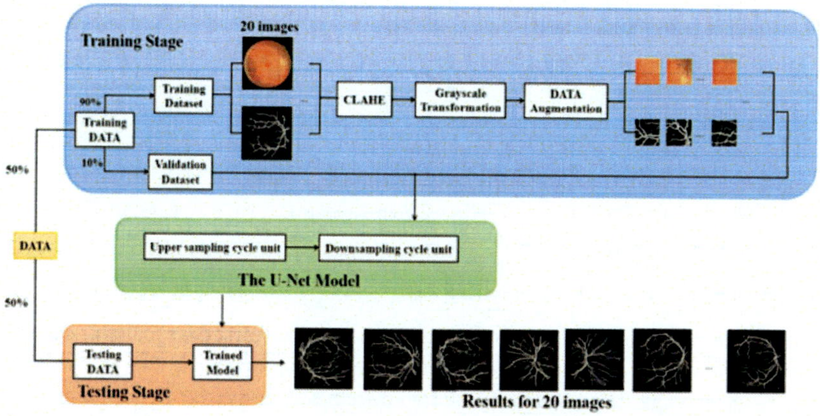

Fig. 13.3 Training process and test results on the DRIVE dataset

these images to obtain a set of 190,000 patches. We divided this dataset in a 9-to-1 ratio for training and validation. The process and consequence of our work are shown in Fig. 13.3.

13.3.3 Evaluation Indexes of the U-Net Model

We use two sets of numerical evaluation metrics to evaluate the performance split outcome of the U-Net method. One group is the standard image segmentation evaluation metrics. The calculated values are shown in Table 13.1. The mathematical expressions of these indicators are as follows:

$$\text{Accuracy} = \frac{TP + TN}{TP + FP + TN + FN} \tag{13.3}$$

$$\text{Specificity} = \frac{TN}{TP + FP} \tag{13.4}$$

$$\text{Recision} = \frac{TP}{TP + FP} \tag{13.5}$$

$$\text{Recall} = \frac{TP}{TP + FN} \tag{13.6}$$

TP, *TN*, *FP*, and *FN* [15] represent true positive, true negative, false positive, and false negative, respectively. Accuracy rate, which is the percentage of right predicted samples within the whole sample; Recall, which expresses the ratio of right predicted positive samples within all positive samples; Precision rate, representing

Table 13.1 Evaluation indexes of the DRIVE databases

Indicators of evaluation	Results of calculation
Jaccard similarity score	0.95158
AUC	0.97380
AUCPR	0.89408
Accuracy	0.95150
Specificity	0.98570
Precision	0.87990

the proportion of positive samples that were correctly predicted out of the total number of optimistic predictions; Specificity, indicating the probability of being rightly predicted in a sample of all negative cases [7].

The ROC curve [7] describes the balance between the true positives and false positives of the classifier. The superior in the performance of the algorithm segmentation, the higher the value, as shown in Fig. 13.4a.

Precision is the abscissa in a P-R curve [7], and Recall is the ordinate. Contrary to the ROC curve, the segmentation performance of the algorithm becomes better as the precision value decreases, as shown in Fig. 13.4b.

In addition, we compare the actual value maps and segmentation maps of fundus images under the DRIVE dataset, and some of the results are shown in Fig. 13.5. We can conclude subjective evaluation that the majority of retinal vessels were accurately segmented in these fundus images, except for some tiny vessels that were not segmented. It shows that the U-Net model trained under this dataset has an advanced performance for vessel segmentation.

Apart from the above indicators, we have also calculated three evaluation indexes for medical image segmentation to estimate the performance of the U-Net model. As shown in Fig. 13.6, the three indexes are introduced as follows.

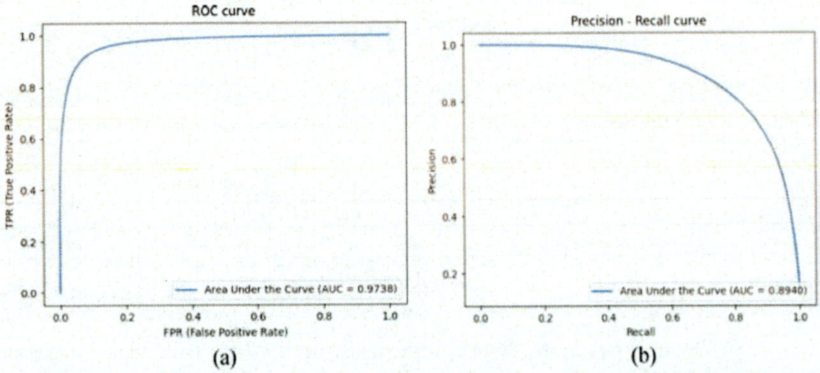

Fig. 13.4 **a** ROC curve of U-Net retinal vessel segmentation based on DRIVE databases. **b** P-R curve of retinal vessel segmentation of U-Net model obtained by training under DRIVE databases

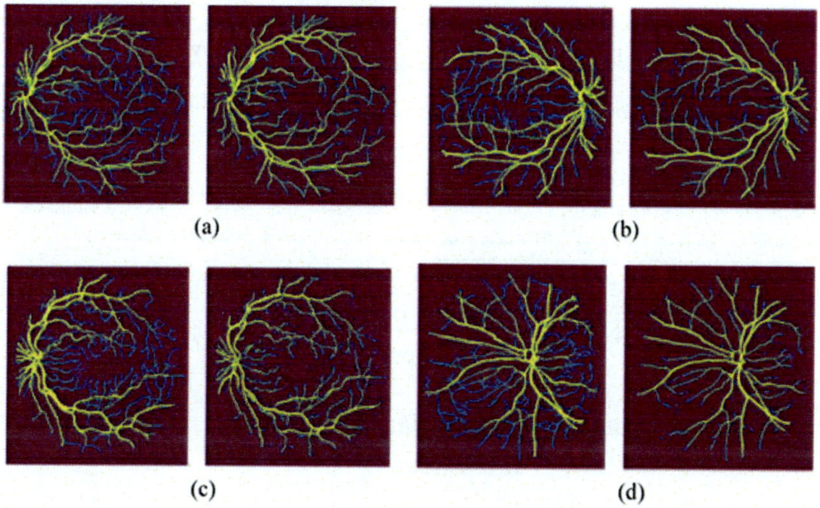

Fig. 13.5 Fundus image truth graph (left) and segmentation graph (right) under DRIVE databases

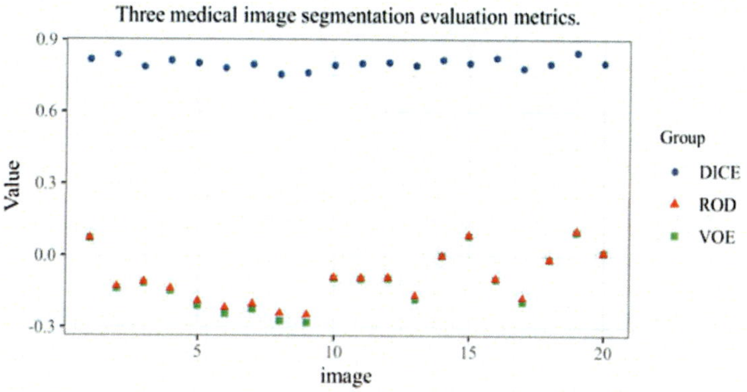

Fig. 13.6 Three medical image segmentation evaluation metrics

The DICE [16] coefficient is a similarity evaluation function. This indicator is widely applied to 3D medical image segmentation, and its value is usually between 0 and 1. The higher its value, the better the model is constructed. The following formula is:

$$\mathrm{DICE} = \frac{2TP}{2TP + FP + FN} \tag{13.7}$$

VOE [17] can be called the volume overlap error, which represents the error rate. The details are as follows:

$$\text{VOE} = 1 - \frac{TP}{TP + FP + FN} \tag{13.8}$$

RVD [17] represents the difference between the volumes of the two, mathematically defined as follows:

$$\text{RVD} = 1 - \frac{FP}{TP + FN} \tag{13.9}$$

According to the results above, we can draw the conclusion that the segmentation performance of the U-Net model is excellent.

13.4 Predictive Generation of Fundus Vessel Segmentation Images

The predicted and original images are in the private dataset for visual comparison. It is observed that the U-Net model is more robust for feature extraction of coarse vessels, but at the same time, more information is obtained on tiny vessels. The above results are displayed in Fig. 13.7.

13.5 Conclusion

This study uses the U-Net model trained under the DRIVE dataset to implement the segmentation of fundus image vessels in the private dataset. It is better for the subsequent work of our private dataset. Moreover, the algorithm decreases the sophistication of the network and improves the segmentation accuracy. The results indicate that our trained resulting U-Net model's area under the curve, accuracy, and specificity are 97.38%, 95.15%, and 98.57%, respectively.

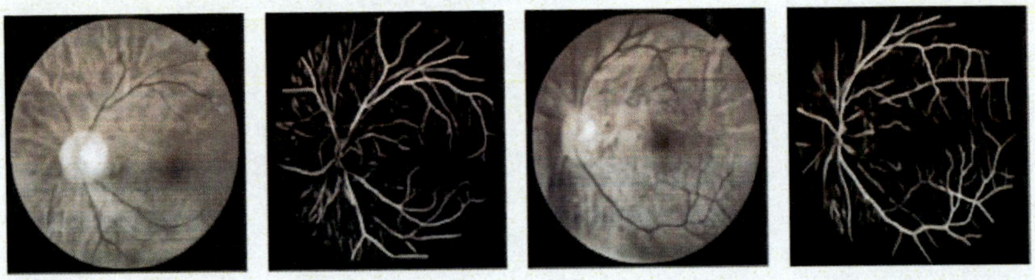

Fig. 13.7 Comparison between the private databases' original image (left) and the predicted image (right)

References

1. Almotiri, J., Elleithy, K., Elleithy, A.: Retinal vessels segmentation techniques and algorithms: a survey. Appl. Sci. **8**(2), 155 (2018)
2. Srinidhi, C.L., Aparna, P., Rajan, J.: Recent advancements in retinal vessel segmentation. J. Med. Syst. **41**(4), 1–22 (2017)
3. Srinivas, S., Sarvadevabhatla, R.K., Mopuri, K.R., Prabhu, N., Kruthiventi, S.S., Babu, R.V.: A taxonomy of deep convolutional neural nets for computer vision. Front. Robot. AI **2**, 36 (2016)
4. Litjens, G., Kooi, T., Bejnordi, B.E., Setio, A.A.A., Ciompi, F., Ghafoorian, M., van der Laak Jeroen, A.W.M., van Ginneken, B., Sánchez, C.I.: A survey on deep learning in medical image analysis. Med. Image Anal. **42**, 60–88 (2017)
5. Xiancheng, W., Wei, L., Bingyi, M., He, J., Jiang, Z., Xu, W., Ji, Z., Hong, G., Zhaomeng, S.: Retina blood vessel segmentation using a U-Net based convolutional neural network. In: Procedia Computer Science: International Conference on Data Science (ICDS 2018), pp. 8–9 (2018).\
6. Gao, X., Cai, Y., Qiu, C., Cui, Y.: Retinal blood vessel segmentation based on the Gaussian matched filter and U-net. In: 2017 10th International Congress on Image and Signal Processing, BioMedical Engineering and Informatics (CISP-BMEI), pp. 1–5. IEEE (2017)
7. Pan, X.Q., Zhang, Q.R., Zhang, H., Li, S.M.: A fundus retinal vessels segmentation scheme based on the improved deep learning U-Net model. IEEE Access **7**, 122634–122643 (2019)
8. Ming, L.L., Qi, X.S.: Improved U-Net fundus retinal vessels segmentation. Appl. Res. Comput. **37**(4), 1–6 (2019)
9. https://ieeexplore.ieee.org/abstract/document/1282003. Accessed 21 Dec 2022
10. Ronneberger, O., Fischer, P., Brox, T.: U-net: convolutional networks for biomedical image segmentation. In: International Conference on Medical image computing and computer-assisted intervention, pp. 234–241. Springer, Cham (2015)
11. https://www.mayiwenku.com/p-27582017.html9. Accessed 21 Dec 2022
12. https://www.modb.pro/db/53049512. Accessed 21 Dec 2022
13. https://blog.csdn.net/fafagege11520/article/details/114287978. Accessed 21 Dec 2022
14. https://blog.csdn.net/opencv_fjc/article/details/10566541014. Accessed 21 Dec 2022
15. Walter, T., Klein, J.C., Massin, P., Erginay, A.: A contribution of image processing to the diagnosis of diabetic retinopathy-detection of exudates in color fundus images of the human retina. IEEE Trans. Med. Imaging **21**, 1236–1243 (2002)
16. Taha, A.A., Hanbury, A.: Metrics for evaluating 3D medical image segmentation: analysis, selection, and tool. BMC Med. Imaging **15**(1), 1–28 (2015)
17. Budak, Ü., Guo, Y., Tanyildizi, E., Şengür, A.: Cascaded deep convolutional encoder-decoder neural networks for efficient liver tumor segmentation. Med. Hypotheses **134**, 109431 (2020)

Chapter 14
Design of License Plate Recognition System Based on OpenCV

Yajuan Wang and Zaiqing Chen

Abstract License plate recognition has become an important part of the transportation system. A system is designed to recognize vehicle images by license plate recognition, letter separation, and recognition letter separation. The first part of the system is the selection of licenses. There are many ways to choose a license plate. After the flag is determined, the character split is performed. Character recognition is the last part, and the recognition method is usually template matching. Segment identifiers match characters in the template library, and the best-matched template is used to accomplish character recognition. A license plate recognition method based on edge recognition and color is proposed to achieve a higher license plate recognition rate. License plate recognition systems use simpler images than those captured by cameras or video.

14.1 Introduction

Because the number of cars is increasing. With the advanced technology to solve the road traffic problem, vehicle monitoring has become the main subject of many research institutions. A driver's license reflects a vehicle's "identity" and provides basic support for the development of an intelligent vehicle management system.

According to the characteristics of license plates, local scientists have proposed a variety of license plate recognition methods [1]. However, many license plate recognition systems are very demanding [2]. Many scientists have studied the license plate color recognition methods and obtained several license plate color recognition algorithms [3]. Subsequently, an identifier location method based on identifier structure was proposed [4]. The system combines edge detection methods to recognize label

Y. Wang · Z. Chen (✉)
School of Information Science and Technology, Yunnan Normal University, Kunming, Yunnan, China
e-mail: Zaiqingchen@ynnu.edu.cn

Z. Chen
Yunnan Key Laboratory of Optoelectronic Information Technology, Kunming, Yunnan, China

S. Patnaik et al. (eds.), *3D Imaging—Multidimensional Signal Processing and Deep Learning*, Smart Innovation, Systems and Technologies 348,
https://doi.org/10.1007/978-981-99-1145-5_14

colors and store them in a collection. Then support vector machine (SVM) is used to filter the collected license plates and select the correct license plate image to display the successful location. The successful positioning of characters in symbol image segmentation includes using multi-character correction method to correct the image before character segmentation, using support vector machine to complete character segmentation, and finally classifying and recognizing the characters correctly.

14.2 Experimental Principle

14.2.1 License Plate Location Method Based on License Plate Color

National license styles are usually blue, white, yellow, and black. Now, the license plates of new energy vehicles are green with black text on them. Based on this function, scientists have proposed a detection method based on license plate color. Figure 14.1 shows the general process of this approach.

Some scientists have proposed a method to convert RGB color space to HSV color space. HSV spaces describe color changes better than RGB spaces. It consists of a composite hue (H), saturation (S), and brightness (V) to describe the name of the color. The HSV color space model is shown in Fig. 14.2.

The steps to locate license plates according to the color of license plates: (1) First, input color images into the system, calculate these using the following Formulas 14.1–14.3, and then use the histogram to compensate and complete the pretreatment.

Fig. 14.1 Locating the license plate based on its color

Fig. 14.2 HSV color space model

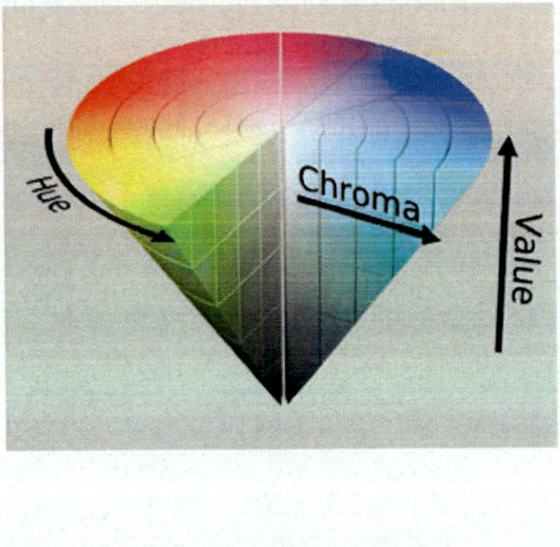

$$h = \begin{cases} 0^{\circ}, \Delta = 0 \\ 60^{\circ} \times \frac{G' - B'}{\Delta} mod6, C_{\max} = R', \\ 60^{\circ} \times \frac{B' - R'}{\Delta} + 2, C_{\max} = B' \\ 60^{\circ} \times \frac{R' - G'}{\max - \min} + 4, C_{\max} = G' \end{cases} \tag{14.1}$$

$$s = \begin{cases} 0, C_{\max} = 0 \\ \frac{\Delta}{C_{\max}}, C_{\max} \neq 0 \end{cases} \tag{14.2}$$

$$v = C_{\max} \tag{14.3}$$

where $R' = R/255$; $G' = G/255$ $B' = B/255$; $C_{\max} = \max(R', G', B')$; $C_{\min} = \min(R', G', B')$; $\Delta = C_{\max} - C_{\min}$.

Pixels across a color image. If the value of the H component in the HSV color space is between 200 and 280, and the value of the S component is between 0.35 and 1.0, it is marked white, and vice versa [5].

(3) The image is binarized and closed, and the outer rectangle of the label is cut to remove the small black dot from the image. This method completes the location of the blue license. If the value of the H color component changes, it can be restored by identifying markers of other colors.

14.2.2 License Plate Location Method Based on Edge Detection

In order to capture the obvious features of license plate edge information, a license plate location method based on edge detection was proposed. There are several edge detection methods, such as Roberts, Prewitt and Sobel [6]. These operators determine the edge of the image according to the feature of the significant change of the gray level of the object edge. The license plate location based on edge detection has the characteristics of short running time, strong noise reduction ability, and fast location speed, including multiple license plate images. However, if the tag is heavily hidden, the position will fail because neither side of the text line is recognized.

In addition, if the license plate is faulty or tilted, the location area is slightly larger than the license plate [7]. Figure 14.3 shows the entire process.

14.2.3 License Plate Correction Methods

Hough algorithm uses the angle corresponding to the maximum repetition point of the parameter to correct the image tilt [8]. His main idea was to convert the image from a variable space to a parameter space that generates many peaks. Peak detection indicates the existence of a straight line or curve. The method is mainly divided into

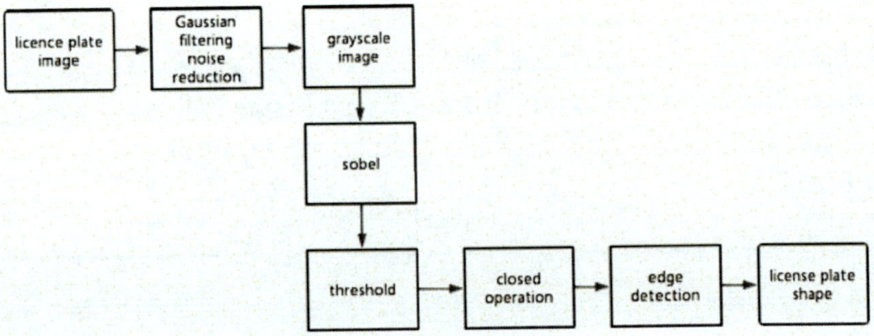

Fig. 14.3 License plate location flowchart based on edge detection

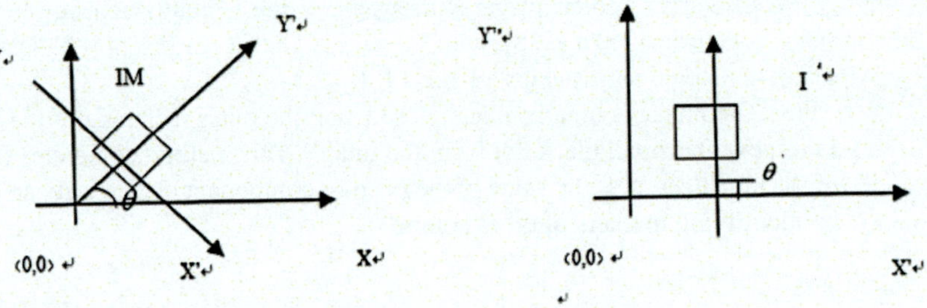

Fig. 14.4 Diagram of randon

two steps: (1) detect the tilt of the support plate and (2) calibrate license plate. Randon calculates the orientation of the image by defining the angle of overlap of the image and obtaining a maximum projection value to determine the angle of tilt of the image. Figure 14.4 shows the steps.

14.2.4 Character Recognition Algorithm Based on Template Matching

Template matching is the comparison of the degree of matching between split characters and default characters in the template library [9]. The best matching character is the recognized character. A common step in pattern matching is to standardize the size of matching character images and template characters and calculate correlations. The correlation operator $R(x, y)$ is represented in Formula 14.4.

$$R(x, y) = \frac{\sum_{m=1}^{M} \sum_{n=1}^{N} F_{xy}(m, n) \times T(m, n)}{\sqrt{\sum_{m=1}^{M} \sum_{n=1}^{N} (F_{xy}(m, n))^2} \sqrt{\sum_{m=1}^{M} \sum_{n=1}^{N} (T(m, n))^2}} \tag{14.4}$$

The character image to be tested is $F(m, n)$, and the submap to be recognized is $F_{xy}(m, n)$. The template image is $T(m, n)$.

The original image is a binary image, and the above equation can be simplified as Formula 14.5, as shown below:

$$D(x, y) = \sum_{m=1}^{M} \sum_{n=1}^{N} F_{xy}(m, n) \oplus T(m, n) \tag{14.5}$$

The result is that the smaller the value of $D(x, y)$, the better the recognition effect.

14.3 Implementation and Results

14.3.1 License Plate Positioning Based on License Plate Color

The system uses a color location license to separate the three components of the HSV by Formulas 14.1–14.3 to calculate. The specific process of license plate color recognition method is as follows: Fig. 14.5a shows the original image of imported vehicles.

As shown in Fig. 14.5b, the transformed image is set from the RGB color space to the black pixel in the HSV color space, similar to the background color and license plate threshold and the white pixel with a smaller threshold. You will then get an image of the license threshold in Fig. 14.6.

Close the license threshold image to get the rectangular connection area. The area of the rectangular connector is completed in Fig. 14.7. Finally, color alignment is accomplished by splitting the label outline using a rectangular combination area, in Fig. 14.8.

Fig. 14.5 Vehicle picture: **a** the original image of the vehicle in RGB space and **b** the RGB color space is converted to HSV color space

Fig. 14.6 License plate threshold image

Fig. 14.7 Closed operation image

Fig. 14.8 Color location license plate image

14.3.2 License Plate Location Based on License Plate Edge Detection

The edge of the license plate is very rich in information, the characteristics of the corresponding pretreatment are more obvious, and here is the processing process. Figure 14.9 shows the license plate picture.

By reducing Gaussian noise, shown in Fig. 14.10, the image in Fig. 14.9 can convert the reduced Gaussian noise into a grayscale image. The result is shown in Fig. 14.11. Edge detection is then performed to highlight the label's edge information. Figure 14.12 shows the result. After an edge is detected, the threshold (binarization) continues to extract the edge to show a distinct black and white effect. Figure 14.13 is the graph representing the result.

When the threshold starts, the extension of the adjacent white area is turned off and connected to the entire area to maintain the operation of the connected area. Figure 14.14 is the figure that shows the result.

Fig. 14.9 License plate

Fig. 14.10 Image after Gaussian noise reduction

Fig. 14.11 Grayscale image

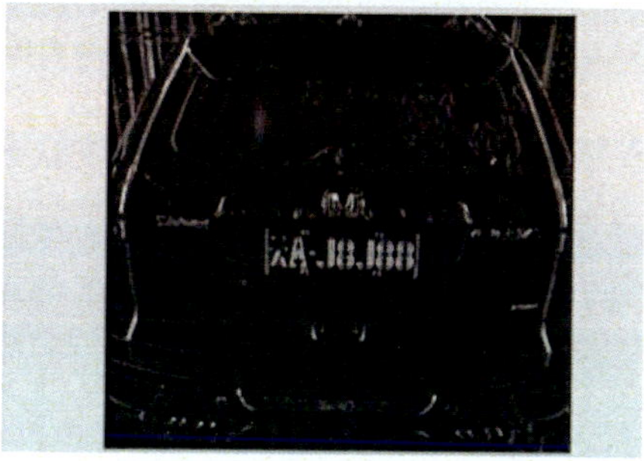

Fig. 14.12 Sobel edge detection

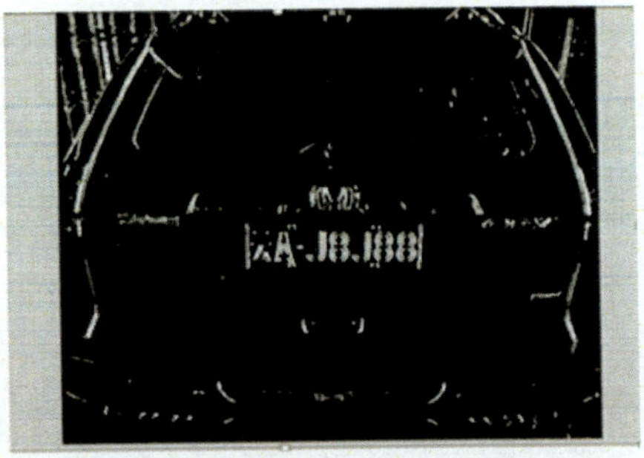

Fig. 14.13 Threshold image

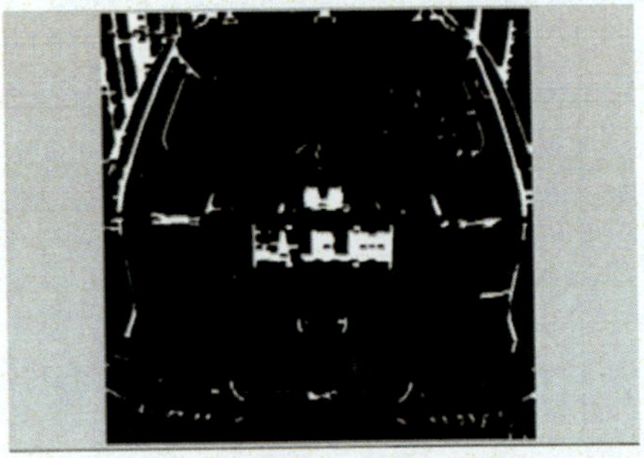

Fig. 14.14 Closed image

Fig. 14.15 Locating the license plate

After completing the previous noise reduction step, place the plate with the initial filter and plate tilt compensation. The license image obtained using this method is shown in Fig. 14.15.

14.3.3 Character Segmentation Method Based on Projection

The loaded license plate image is shown in Fig. 14.16: The loaded license plate image is grayscale processed, and the grayscale license plate image is shown in Fig. 14.17.

Fig. 14.16 License plate image

Fig. 14.17 License plate image after grayscale

Fig. 14.18 License plate image after threshold

Fig. 14.19 Horizontal projection without interference

By distorting the gray license plate image, the effect of black and white license plate is obvious, and the edge information is also obvious. Figure 14.18 shows an image of the license threshold. However, the nameplate has four fixed digit pegs at the top and bottom that affect text segmentation. Therefore, the interference of the claw is eliminated by horizontal projection, and the license plate image in Fig. 14.19 is obtained.

By eliminating the interference of the nails, you will notice that the image on the license plate is clearer and the text is also split. Since the connection field cannot distinguish the structure of the kanji, the first non-kanji on the license plate must first be found, and then the forward movement range must be defined to cut the kanji. Figures 14.20 and 14.21 show the registered Chinese characters, text, and numbers.

Fig. 14.20 Chinese characters

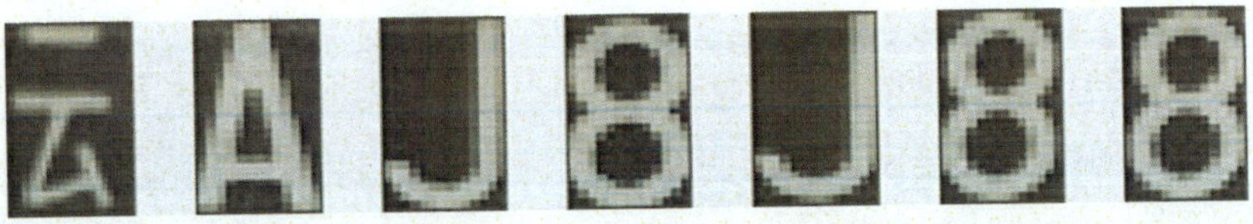

Fig. 14.21 Split characters

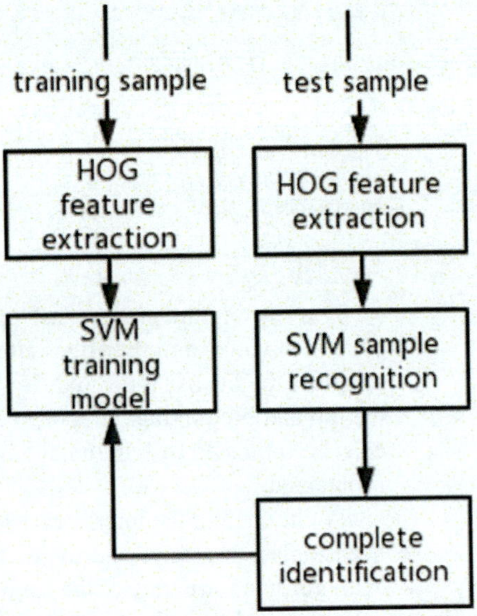

Fig. 14.22 SVM identification characters

14.3.4 SVM-Based Character Recognition Method

At present, multi-recognition support vector machines are mainly divided into two categories: (1) several ideas and (2) several combinations of binary classifiers [10]. The character recognition method used in this chapter is a fast and high detection rate SVM machine learning method. Figure 14.22 shows the main steps. First, the text image is converted into feature vector by Hog feature extraction method, and the feature vector is used for classification.

14.4 Conclusion

This paper not only describes the accurate location and recognition method of license plate, but also realizes a license plate location method combining color and edge recognition of license plate, which provides a reference for the future development of vehicle management systems.

References

1. Chamadiya, M.A.H., Chaudhary, M.D., Ramana, T.V.: An expanded-haar wavelet transform and morphological deal based approach for vehicle license plate localization in Indian conditions. Int. J. Res. Eng. Technol. **3**(04), 685–692 (2014)
2. Liu, W.J., Jiang, Q.L., Zhang, C.: License plate location algorithm based on CNN color image edge detection. Acta Autom. Sin. **35**(12), 1503–1512 (2009). (In Chinese)
3. Li, W.J., Liang, D.Q., Zhang, Q., et al.: A new method for license plate localization based on edge color pairs. Chin. J. Comput. **27**(2), 204–208 (2004). (In Chinese)
4. Lin, W., Du, H.: A license plate location method combining the license plate location method based on the HSV color space and the license plate location method based on the boundary detection. In: Proceedings of 2016 4th International Conference on Machinery, Materials and Computing Technology (ICMMCT2016), pp. 1940–1943 (2016)
5. Li, X.F.: Research on tilt license plate correction and recognition based on deep learning. Tianjin Vocational and Technical Normal University (2019). (In Chinese)
6. Yan, Q.: Analysis of common license plate location algorithm. Microcomput. Appl. **02**, 5–7 (2010). (In Chinese)
7. Liang, J.: Research on license plate image location algorithm based on projection template method. Chongqing University (2010). (In Chinese)
8. Hu, F.F.: Research on license plate location segmentation algorithm. Kunming University of Science and Technology (2011). (In Chinese)
9. Li, Y.H.: Tilt image correction technology based on Hough transform. J. Hunan Inst. Technol. (Nat. Sci. Ed.) **3** (2019). (In Chinese)
10. Li, F.X., Wang, X., Zhang, C., Zhou, M.: Support vector machine classification algorithm based on boundary points. J. Shaanxi Univ. Technol. (Nat. Sci. Ed.) **38**(03), 30–38 (2022). (In Chinese)

Chapter 15
Traveling Wave Solutions of the Nonlinear Gardner Equation with Variable-Coefficients Arising in Stratified Fluids

Qian Wang and Guohong Liang

Abstract This paper aims to transform the nonlinear development equation into elementary integral form by the trial equation method. The Gardner equation which can be used to describe stratified fluids is studied. By applying the trial equation method, 47 types of traveling wave solutions are obtained. The method is applied in the generalized Gardner equation with variable coefficients. It shows that the method is very effective for solving some linear or simple nonlinear partial differential equations with variable coefficients.

15.1 Introduction

Many mathematical models in physics and engineering are ultimately reduced to the variable-coefficient partial differential equations (PDEs). To seek the exact solution of PDEs with variable coefficients is an important aspect of scientific research. In recent years, people put forward various kinds of methods to find the exact solution, for example, Darboux transformation [1], Bäcklund transformation method [2–4], inverse scattering transformation approach [5], Lie group analysis method [6–9], etc. The trial equation method [10–13] is an effective finding nonlinear PDEs including variable-coefficients equations.

Gardner equation is commonly used in fluid physics, quantum field theory, plasma physics, and other branches of physical science [14–16]. It is also used to describe wave phenomena in solids plasmas [17–20].

In this paper, the one-dimensional variable-coefficient Gardner equation [21–24] is considered. This is the result of

$$U_t + A(t)\,UU_x + C(t)\,U^2U\_x + B(t)\,U_{xxx} = 0 \tag{15.1}$$

Q. Wang (✉) · G. Liang
Foundamental Department of Air Force Engineering University, Xi′an 710051, Shaanxi, China
e-mail: wangqian5444@163.com

S. Patnaik et al. (eds.), *3D Imaging—Multidimensional Signal Processing and Deep Learning*, Smart Innovation, Systems and Technologies 348,
https://doi.org/10.1007/978-981-99-1145-5_15

where $U(x,t)$ is the function with x and t, $U_x = \frac{\partial U}{\partial x}$, $U_{xxx} = \frac{\partial^3 U}{\partial x^3}$, $A(t)A(t)$, $B(t)$, $C(t)$ are smooth nonzero functions with respect to t.

Equation (15.1), also known as the combined mKdV KdV equation, is derived from the study of the effect of surface tension on solitary waves in stratified fluid [19, 20]. The exact solutions of Eq. (15.1) are obtained by the improved tanh-coth method in Ref. [21]. And Lie symmetries of Eq. (15.1) were studied in Ref. [22].

The rest of this article is arranged as follows. In Sect. 15.2, we study the Gardner equation with variable-coefficients applying trial equation method. In Sect. 15.3, 47 types of traveling wave solutions are obtained by applying trial equation method. Finally, the conclusions are given.

15.2 Application of Trial Equation Method

In view of trial equation method, we first perform the traveling wave transformation

$$U(x,\lambda) = U(\beta), \quad \beta = k(\lambda)x + \omega(\lambda) \tag{15.2}$$

where $k(\lambda)$ and $\omega(\lambda)$ are undetermined time-dependent parameters.

Substituting (15.2) into Eq. (15.1), we get

$$(k'(\lambda)x + \omega'(\lambda))U' + A(\lambda)k(\lambda)UU' + C(\lambda)k(\lambda)U^2U' + B(\lambda)k^3(\lambda)U' = 0 \tag{15.3}$$

where prime expresses differentiation with respect to β.

Now assume that the exact solution of Eq. (15.1) satisfies the following trial equation [9–14]

$$(U')^2 = \sum_{i=0}^{n} v_i U_i^i \tag{15.4}$$

where $v_i (i = 0, 1, \ldots, n)$ are constants and n is the integer to be determined later.

Balancing terms u' and u^2u' in Eq. (15.3) yields $n = 4$. Therefore, we obtain

$$(U')^2 = v_0 + v_1U + v_2U^2 + v_3U^3 + v_4U^4 \tag{15.5}$$

From (15.5), we have

$$U' = (v_2 + 3v_3U + 6v_4U^2)U' \tag{15.6}$$

Substituting (15.6) into Eq. (15.3) and letting the coefficients of u^i to zero, we obtain

$$k'(\lambda)x + \omega' + v_2 B(\lambda)k^3(\lambda) = 0 \tag{15.7}$$

$$A(\lambda)kx + 3v_3 B(\lambda)k^3 = 0 \tag{15.8}$$

$$C(\lambda)k(\lambda)x + 6v_3 B(\lambda)k^3(\lambda) = 0 \tag{15.9}$$

After solving Eqs. (15.7)–(15.9) is obtained

$$k\,(\lambda) = k \tag{15.10}$$

$$\omega'(\lambda) + v_2 B(\lambda)k^3(\lambda) = 0 \tag{15.11}$$

$$A(\lambda) = -3v_3 k^2(\lambda)B(\lambda) \tag{15.12}$$

$$C(\lambda) = -6v_4 k^2(\lambda)B(\lambda) \tag{15.13}$$

where k, v_3, v_4 are nonzero constants.

If $v_2 = 0, \omega(\lambda) = c$ is a nonzero constant.

If $v_2 \neq 0, \omega(\lambda) = -v_2 k^3 \int B(\lambda)d\lambda$.

15.3 Exact Solutions of Eq. (15.1)

Next, we give the exact solution of Eq. (15.1). Obviously, the key is to find out the solutions of Eq. (15.5). The solutions of Eq. (15.5) were studied in some literatures [25–27]. Here we briefly introduce the following general elliptic Eq. (15.14)

$$\left(\frac{d\varphi(\beta)}{d\beta}\right)^2 = v_0 + v_1\varphi(\beta) + v_2\varphi^2(\beta) + v_3\varphi^3(\beta) + v_4\varphi^4(\beta) \tag{15.14}$$

where $\varphi(\beta)$ is the function with respect to β, and v_0, v_1, v_2, v_3, v_4 are constants.

In the literatures [26–28], the exact solutions of some special cases for Eq. (15.14) are given.

When $v_0 \neq 0, v_1 \neq 0, v_2 \neq 0, v_3 \neq 0, v_4 \neq 0$, three parameters r, p, s may exist, such that

$$\begin{aligned}\left(\frac{d\varphi(\beta)}{d\beta}\right)^2 &= v_0 + v_1\varphi(\beta) + g_2\varphi^2(\beta) + v_3\varphi^3(\beta) + v_4\varphi^4(\beta)\\ &= (r + p\varphi(\beta) + s\varphi^2(\beta))^2\end{aligned} \tag{15.15}$$

If the following relations hold, then Eq. (15.15) is satisfied:

$$v_0 = r^2, v_1 = 2rp, v_2 = 2rp + p^2, v_3 = 2ps, v_4 = s^2.$$

When $v_2 = 0, v_0 \neq 0, v_1 \neq 0, v_3 \neq 0, v_4 \neq 0$, three parameters r, p, s may exist, that make

$$\left(\frac{d\varphi(\beta)}{d\beta}\right)^2 = v_0 + v_1\varphi(\beta) + v_3\varphi^3(\beta) + v_4\varphi^4(\beta) = (r + p\,\varphi(\beta) + s\,\varphi^2(\beta))^2 \tag{15.16}$$

When $v_0 = v_1 = 0, v_2 \neq 0, v_3 \neq 0, v_4 \neq 0$, Eq. (15.14) is reduced into the following

$$\left(\frac{d\varphi}{d\beta}\right)^2 = v_2\varphi^2(\beta) + v_3\varphi^3(\beta) + v_4\varphi^4(\beta) \tag{15.17}$$

According to exact solutions of the general elliptic Eq. (15.15)–(15.17), we can obtain the traveling wave solutions of variable-coefficient Gardner Eq. (15.1).

Case 1: If $v_0 \neq 0, v_1 \neq 0, v_2 \neq 0, v_3 \neq 0, v_4 \neq 0$, and $\beta = kx - v_2k^3 \int B(\lambda)d\lambda$ the following solutions of Eq. (15.1) are listed:

Type 1. When $p^2 - 4rs > 0$ and $sp \neq 0$(or $rs \neq 0$),

$$u_1(x,\lambda) = -\frac{1}{2s}(p + \sqrt{p^2 - 4rs}\tan v(\frac{\sqrt{p^2 - 4rs}}{2}\beta)),$$

$$u_2(x,\lambda) = -\frac{1}{2s}(p + \sqrt{p^2 - 4rs}\cot v(\frac{\sqrt{p^2 - 4rs}}{2}\beta)),$$

$$u_3(x,t) = -\frac{1}{2s}(p + \sqrt{p^2 - 4rs}(\tan v(\sqrt{p^2 - 4rs}\beta) \pm i\sec v(\sqrt{p^2 - 4rs}\beta))),$$

$$u_4(x,\lambda) = -\frac{1}{2s}(p + \sqrt{p^2 - 4rs}(\cot v(\sqrt{p^2 - 4rs}\beta) \pm \csc v(\sqrt{p^2 - 4rs}\beta))),$$

$$u_5(x,\lambda) = -\frac{1}{4s}(2p + \sqrt{p^2 - 4rs}(\tan v(\frac{\sqrt{p^2 - 4rs}}{4}\beta) + \cot v(\frac{\sqrt{p^2 - 4rs}}{4}\beta))),$$

$$u_6(x,\lambda) = \frac{1}{2s}(-p + \frac{\sqrt{(E^2 + F^2)(p^2 - 4rs)} - E\sqrt{p^2 - 4rs}\cos v(\sqrt{p^2 - 4rs}\beta)}{E\sin v(\sqrt{p^2 - 4rs}\beta) + F}),$$

$$u_7(x,\lambda) = \frac{1}{2s}(-p - \frac{\sqrt{(F^2 - E^2)(p^2 - 4rs)} + E\sqrt{p^2 - 4rs}\sin v(\sqrt{p^2 - 4rs}\beta)}{E\cos v(\sqrt{p^2 - 4rs}\beta) + F}),$$

where E and F are nonzero real constants which satisfy $F^2 - E^2 > 0$;

$$u_8(x,\lambda) = \frac{2r\cos v(\frac{\sqrt{p^2-4rs}}{2}\beta)}{\sqrt{p^2 - 4rs}\sin v(\frac{\sqrt{p^2-4rs}}{2}\beta) - p\cos v(\frac{\sqrt{p^2-4rs}}{2}\beta)},$$

$$u_9(x,\lambda)=\frac{2r\sin v(\frac{\sqrt{p^2-4rs}}{2}\beta)}{p\sin v(\frac{\sqrt{p^2-4rs}}{2}\beta)-\sqrt{p^2-4rs}\cos v(\frac{\sqrt{p^2-4rs}}{2}\beta)},$$

$$u_{10}(x,\lambda)=\frac{2r\cos v(\sqrt{p^2-4rs}\beta)}{\sqrt{p^2-4rs}\sin v(\sqrt{p^2-4rs}\beta)-p\cos v(\sqrt{p^2-4rs}\beta)\pm i\sqrt{p^2-4rs}},$$

$$u_{11}(x,\lambda)=\frac{2r\sin v(\sqrt{p^2-4rs}\beta)}{-psinv(\sqrt{p^2-4rs}\beta)+\sqrt{p^2-4rs}\cos v(\sqrt{p^2-4rs}\beta)\pm\sqrt{p^2-4rs}},$$

$$u_{12}(x,\lambda)$$

$$=\frac{4r\cos v(\frac{\sqrt{p^2-4rs}}{4}\beta)\sin v(\frac{\sqrt{p^2-4rs}}{4}\beta)}{-2p\sin v(\frac{\sqrt{p^2-4rs\beta}}{4})\cos v(\frac{\sqrt{p^2-4rs\beta}}{4})+2\sqrt{p^2-4rs}\cos v^2(\frac{\sqrt{p^2-4rs\beta}}{4})-\sqrt{p^2-4rs}}.$$

Type 2. When $p^2-4sr<0$ and $ps\neq 0$ (or $sr\neq 0$),

$$u_{13}(x,\lambda)=\frac{1}{2s}(-p+\sqrt{4rs-p^2}\tan(\frac{\sqrt{4rs-p^2}}{2}\beta)),$$

$$u_{14}(x,\lambda)=-\frac{1}{2s}(p+\sqrt{4rs-p^2}\cot(\frac{\sqrt{4rs-p^2}}{2}\beta)),$$

$$u_{15}(x,\lambda)=\frac{1}{2s}(-p+\sqrt{4rs-p^2}(\tan(\sqrt{4rs-p^2}\beta)\pm\sec(\sqrt{4rs-p^2}\beta))),$$

$$u_{16}(x,\lambda)=-\frac{1}{2s}(p+\sqrt{4rs-p^2}(\cot(\sqrt{4rs-p^2}\beta)\pm\csc(\sqrt{4rs-p^2}\beta))),$$

$$u_{17}(x,\lambda)=\frac{1}{4s}(-2p+\sqrt{4rs-p^2}(\tan(\frac{\sqrt{4rs-p^2}}{4}\beta)-\cot(\frac{\sqrt{4rs-p^2}}{4}\beta))),$$

$$u_{18}(x,\lambda)=\frac{1}{2s}(-p+\frac{\pm\sqrt{(E^2-F^2)(4rs-p^2)}-E\sqrt{4rs-p^2}\cos(\sqrt{4qr-p^2\beta})}{E\sin(\sqrt{4rs-p^2\beta})+F}),$$

$$u_{19}(x,\lambda)=\frac{1}{2s}(-p-\frac{\pm\sqrt{(E^2-F^2)(4rs-p^2)}-E\sqrt{4rs-p^2}\sin(\sqrt{4rs-p^2\beta})}{E\cos(\sqrt{4rs-p^2\beta})+F}),$$

where E and F are nonzero real constants which satisfy $E^2-F^2>0$;

$$u_{20}(x,\lambda)=\frac{2r\cos(\frac{\sqrt{4rs-p^2}}{2}\beta)}{\sqrt{4rs-p^2}\sin(\frac{\sqrt{4rs-p^2}}{2}\beta)+p\cos(\frac{\sqrt{4rs-p^2}}{2}\beta)},$$

$$u_{21}(x,\lambda)=\frac{2r\sin(\frac{\sqrt{4rs-p^2}}{2}\beta)}{-p\sin(\frac{\sqrt{4rs-p^2}}{2}\beta)+\sqrt{4rs-p^2}\cos(\frac{\sqrt{4rs-p^2}}{2}\beta)},$$

$$u_{22}(x,\lambda)=-\frac{2r\cos(\sqrt{4rs-p^2\beta})}{\sqrt{4rs-p^2}\sin(\sqrt{4rs-p^2\beta})+p\cos(\sqrt{4rs-p^2\beta})\pm\sqrt{4rs-p^2}},$$

$$u_{23}(x,\lambda)=\frac{2r\sin(\sqrt{4rs-p^2}\beta)}{-p\sin(\sqrt{4rs-p^2}\beta)+\sqrt{4rs-p^2}\cos(\sqrt{4rs-p^2}\beta)\pm\sqrt{4rs-p^2}},$$

$$u_{24}(x,\lambda)$$
$$=\frac{4r\cos(\frac{\sqrt{4rs-p^2}}{4}\beta)\sin(\frac{\sqrt{4rs-p^2}}{4}\beta)}{-2p\sin(\frac{\sqrt{4rs-p^2}\beta}{4})\cos(\frac{\sqrt{4rs-p^2}\beta}{4})+2\sqrt{4rs-p^2}\cos^2(\frac{\sqrt{4rs-p^2}\beta}{4})-\sqrt{4rs-p^2}}.$$

Type 3. When $r=0$ and $ps\neq 0$,

$$u_{25}(x,\lambda)=-\frac{pd}{s(d+\cos v(p\beta)-\sin v(p\beta)},$$

$$u_{26}(x,\lambda)=\frac{p(\cos v(p\beta)+\sin v(p\beta))}{s(d+\cos v(p\beta)+\sin v(p\beta))},$$

where d is a constant.

Type 4. When $s\neq 0$ and $r=p=0$,

$$u_{27}(x,\lambda)=\frac{-1}{s\beta+c_1},$$

where $c1$ is a constant.

Case 2 If $v_2=0$, $v_0\neq 0$, $v_1\neq 0$, $v_3\neq 0$, $v_4\neq 0$, $sr<0$ and $\beta=kx+c$, the following solutions of Eq. (15.1) are listed:

$$u_{28}(x,\lambda)=-\frac{1}{2s}(\pm\sqrt{-2rs}+\sqrt{-6rs}\tan v(\frac{\sqrt{-6rs}}{2}\beta)),$$

$$u_{29}(x,\lambda)=-\frac{1}{2s}(\pm\sqrt{-2rs}+\sqrt{-6rs}\cot v(\frac{\sqrt{-6rs}}{2}\beta)),$$

$$u_{30}(x,\lambda)=-\frac{1}{2s}(\pm\sqrt{-2rs}+\sqrt{-6rs}(\tan v(\sqrt{-6rs}\beta)\pm i\sec v(\sqrt{-6rs}\beta))),$$

$$u_{31}(x,\lambda)=-\frac{1}{2s}(\pm\sqrt{-2rs}+\sqrt{-6rs}(\cot v(\sqrt{-6rs}\beta)\pm \csc v(\sqrt{-6rs}\beta))),$$

$$u_{32}(x,\lambda)=-\frac{1}{4s}(\pm 2\sqrt{-2rs}+\sqrt{-6rs}(\tan v(\frac{\sqrt{-6rs}}{4}\beta)+\cot v(\frac{\sqrt{-6rs}}{4}\beta))),$$

$$u_{33}(x,\lambda)=\frac{1}{2s}(\mp\sqrt{-2rs}+\frac{\sqrt{(E^2+F^2)(-6rs)}-E\sqrt{-6rs}\cos v(\sqrt{-6rs}\beta)}{E\sin v(\sqrt{-6rs}\beta)+F}),$$

$$u_{34}(x,\lambda)=\frac{1}{2s}(\mp\sqrt{-2rs}-\frac{\sqrt{(F^2-E^2)(-6rs)}+E\sqrt{-6rs}\sin v(\sqrt{-6rs}\beta)}{E\cos v(\sqrt{-6rs}\beta)+F}),$$

where E and F are nonzero real constants which satisfies $F^2 - E^2 > 0$;

$$u_{35}(x,\lambda) = \frac{2r\cos v(\frac{\sqrt{-6rs}}{2}\beta)}{\sqrt{-6rs}\sin v(\frac{\sqrt{-6rs}}{2}\beta) \mp \sqrt{-2rs}\cos v(\frac{\sqrt{-6rs}}{2}\beta)},$$

$$u_{36}(x,\lambda) = \frac{-2r\sin v(\frac{\sqrt{-6rs}}{2}\beta)}{\pm\sqrt{-2rs}\sin v(\frac{\sqrt{-6rs}}{2}\beta) - \sqrt{-6rs}\cos v(\frac{\sqrt{-6rs}}{2}\beta)},$$

$$u_{37}(x,\lambda) = \frac{2r\cos v(\sqrt{-6rs}\beta)}{\sqrt{-6rs}\sin v(\sqrt{-6rs}\beta) \mp \sqrt{-2rs}\cos v(\sqrt{-6rs}\beta) \pm i\sqrt{-6rs}},$$

$$u_{38}(x,\lambda) = \frac{2r\sin v(\sqrt{-6rs}\beta)}{\mp\sqrt{-2rs}\sin v(\sqrt{-6rs}\beta) + \sqrt{-6rs}\cos v(\sqrt{-6rs}\beta) \pm \sqrt{-6rs}},$$

$$u_{39}(x,\lambda) = \frac{4r\cos v(\frac{\sqrt{-6rs}}{4}\beta)\sin v(\frac{\sqrt{-6rs}}{4}\beta)}{\mp 2\sqrt{-2sr}\sin v(\frac{\sqrt{-6rs}\beta}{4})\cos v(\frac{\sqrt{-6rs}\beta}{4}) + 2\sqrt{-6rs}\cos v^2(\frac{\sqrt{-6rs}\beta}{4}) - \sqrt{-6rs}}.$$

Case 3: If $v_0 = v_1 = 0$, $v_2 \neq 0$, $v_3 \neq 0$, $v_4 \neq 0$, and $\beta = kx - v_2 k^3 \int B(\lambda)d\lambda$, the following solutions of Eq. (15.1) are listed:

Type 1. If $v_2 = 1$, $v_3 = -\frac{2v}{a}$, $v_4 = -\frac{v^2-b^2}{a^2}$, Eq. (15.1) has the solution

$$u_{40}(x,\lambda) = \frac{a\sec v\beta}{b + v\sec v\beta}$$

where a, b, v are constants.

Type 2. If $v_2 = 1$, $v_3 = -\frac{2v}{a}$, $v_4 = -\frac{v^2+b^2}{a^2}$, Eq. (15.1) has the solution

$$u_{41}(x,\lambda) = \frac{a\csc v\beta}{b + \csc v\beta}$$

where a, b, v are constants.

Type 3. If $v_2 = 4$, $v_3 = -\frac{4(2b+d)}{a}$, $v_4 = -\frac{v^2+4b^2+4bd}{a^2}$, Eq. (15.1) has the solution

$$u_{42}(x,\lambda) = \frac{a\sec v^2\beta}{b\sec v^2\beta + v\tan v\beta + d}$$

where a, b, v, d are constants.

Type 4. If $v_2 = 4$, $v_3 = \frac{4(d-2b)}{a}$, $v_4 = -\frac{v^2+4b^2-4bd}{a^2}$, Eq. (15.1) has the solution

$$u_{43}(x,\lambda) = \frac{a\csc v^2\eta}{b\csc v^2\beta + v\cot v\beta + d}$$

where a, b, v, d are constants.

Type 5. If $v_2 = -4$, $v_3 = \frac{4(2b+d)}{a}$, $v_4 = -\frac{v^2+4b^2-4bd}{a^2}$, Eq. (15.1) has the solution

$$u_{44}(x,\lambda) = \frac{a \sec v^2 \beta}{b \sec v^2 \beta + v \tan \beta + d}$$

$$u_{45}(x,\lambda) = \frac{a \csc^2 \beta}{b \csc^2 \beta + v \cot \beta + d}$$

where a, b, v, d are constants.

Type 6. If $v_2 = -1$, $v_3 = \frac{2d}{a}$, $v_4 = -\frac{b^2-d^2}{a^2}$, Eq. (15.1) has the solution

$$u_{46}(x,\lambda) = \frac{a \sec \beta}{b + d \sec \beta}$$

$$u_{47}(x,\lambda) = \frac{a \csc \beta}{b + d \csc \beta}$$

where a, b, d are constants.

15.4 Conclusions

In this paper, the trial equation method is used to construct the exact solutions of Gardner equation with variable coefficients. Then 47 types of exact solutions are obtained, some of which are derived for the first time, such as $u_{37}(x,\lambda)$, $u_{38}(x,\lambda)$, etc. It is complicated to solve the exact solution of the variable coefficient partial differential equation. The trial equation method is relatively effective in solving some linear or simple nonlinear partial differential equations with variable coefficients. So we can deal with some more complex nonlinear partial differential equations with variable coefficients applying trial equation method.

References

1. Matveev, V.B.: Darboux transformation and explicit solutions of the Kadomtcev-Petviaschvily equation, depending on functional parameters. Lett. Math. Phys. **3**, 213–216 (1979)
2. Li, Y., Ma, W.X., Zhang, J.E.: Darboux transformations of classical Boussinesq system and its new solutions. Phys. Lett. A **275**, 60–66 (2000)
3. Vakhnenko, V.O., Parks, E.J.: A Bäcklund transformation and the inverse scattering transform method for the generalized Vakhnenko equation. Chaos Solions Fractals **17**, 683–692 (2003)
4. Hirota, R.: Direct method of finding exact solutions of nonlinear evolution equations. In: Bullough, R., Caudrey, P. (eds.) Bäcklund Transformation. Berloin Springer (1980)
5. Ablowitz, M.J., Segur, H.: Solitons and the inverse scattering transform. Society for Industrial and Applied Mathematics (1981)
6. Ibragimov, N.H.: CRC Handbook of Lie Group Analysis of Differential Equations, vol. 3. CRC Press (1995)
7. Orhan, O., Torrisi, M., Ttacina, R.: Group methods applied to a reaction-diffusion system generalizing Proteus Mirabilis models. Commun. Nonlinear. Sci. 70223–70233 (2019)

8. Liu, C.S.: Trial equation method to solve the exact solutions for two kinds of KdV equations with variable coefficients. Acta Phys. Sin. **54**, 4506–4510 (2005)
9. Liu, C.S.: Trial equation method and its applications to nonlinear evolution equations. Acta Phys. Sin. **54**, 2505–2509 (2005)
10. Liu, C.S.: Trial equation method and its applications. Commun. Theor. Phys. **45**, 395–397 (2006)
11. Ekici, M., Mirzazadeh, M., Sonmezoglu, A., et al.: Optical solitons with anti-cubic nonlinearity by extended trial equation method. Optik **136**, 368–373 (2017)
12. Rui, C., Jian, Z.: Trial function method and exact solutions to the generalized nonlinear Schrödingger equation with time-dependent coefficient. Chin. Phys. B 22100507 (2013)
13. Triki, H., Wazwaz, A.M.: Trial equation method for solving the generalized Fisher equation with variable coefficients. Phys. Lett. A **380**, 1260–1262 (2016)
14. Biswas, A., Yildirim, Y., Yasar, E., et al.: Optical soliton perturbation with full nonlinearity by trial equation method. Optik **157**, 1260–1262 (2016)
15. Fu, Z., Liu, S., Liu, S.: New kinds of solutions to Gardner equation. Chaos Solitons Fractals **20**, 301–309 (2004)
16. Yan, Z.: Jacobi elliptic solutions of nonlinear wave equations via the new sinh-Gordon equation expansion method. MM. Res. **22**, 363–375 (2003)
17. Baldwin, D., Goktas, U., Hereman, W., Hong, L., Martino, R.S., Miller, J.C.: Symbolic computation of exact solutions in hyperbolic and elliptic functions for nonlinear PDEs. J. Symb. Comput. **37**, 669–705 (2004)
18. Hereman, W., Nuseir, A.: Symbolic method to construct exact solutions of nonlinear partial differential equations. Math. Comput. Simul. **43**, 13–27 (1997)
19. Ramollo, M.P.: Internal solitary waves in a two-layer fluid with surface tension. WIT Trans. Eng. Sci. **9** (1970)
20. Molati, M., Ramollo, M.P.: Symmetry classification of the Gardner equation with time-dependent coefficients arising in stratified fluids. Commun. Nonlinear. Sci. **17**, 1542–1548 (2012)
21. Vaneeva, O., Kuriksha, O., Sophocleous, C.: Enhanced group classification of Gardner equations with time-dependent coefficients. Commun. Nonlinear. Sci. **22**, 1243–1251 (2015)
22. Sierra, C.A.G., Molati, M., Ramollo, M.P.: Exact solutions of a generalized KdV-mKdV equation. Int. J. Nonlinear Sci. **13**, 94–98 (2012)
23. De la Rosa, R., Gandarias, M.L., Bruzón, M.S.: Equivalence transformations and conservation laws for a generalized variable-coefficient Gardner equation. Commun. Nonlinear Sci. Numer. Simul. **40**, 71–79 (2016)
24. Wazwaz, A.M.: A study on KdV and Gardner equations with time-dependent coefficients and forcing terms. Appl. Math. Comput. **217**, 2277–2281 (2010)
25. Yomba, E.: The extended Fan's sub-equation method and its application to Kdv-MKdv BKK and variant Boussinesq equations. Phys. Lett. A **336**, 463–476 (2005)
26. Jiong, S.: Auxiliary equation method for solving nonlinear partial differential equations. Phys. Lett. A **309**, 387–396 (2003)
27. Xie, F., Zhang, Y., Lü, Z.: Symbolic computation in non-linear evolution equation: application to $(3+1)$-dimensional Kadomtsev-Petviashvili equation. Chaos Solitons Fractals **24**(1), 257–263 (2005)
28. Fan, E.: Uniformly constructing a series of explicit exact solutions to nonlinear equations in mathematical physics. Chaos Solitons Fractals **16**, 819–839 (2003)

Chapter 16
Research on the Construction of Food Safety Standards Training System Based on 3D Virtual Reality Technology

Peng Liu, Min Duan, Shuang Ren, Shanshan Yuan, Yue Dai, Yiying Nian, and Wen Liu

Abstract This paper adopts technologies such as 3D virtual simulation technology and human–computer interaction technology to develop the foods safety standards interactive question–answering system and intelligent scene-specific foods safety standards training and implementation evaluation platform with regard to different subjects, realizing the diversified training on foods safety standards.

16.1 Introduction

The concept of virtual training was firstly proposed by USA in 1990s. Its appearance attracted the extensive attention of scientific research institutions, military sectors, and business circles of each country worldwide, which was deemed as an effective method to ensure the large-scale production of advanced products. In 2019, Central South University had researched the development of non-coal mine fire emergency training system based on VR technology, established the mine fire virtual scene, and overcome the problem that traditional emergency training method of non-coal mine fire was not good and it was hard to realize partial training contents [1]; Liu Dunwen et al. had utilized the virtual reality technology to develop the tunnel fire emergency training system, realized the fire emergency training for tunnel rescue personnel and management personnel in the virtual reality environment of high immersion, improved the enthusiasm of personnel to participate in the emergency rescue drills, and reduced the costs of emergency drills [2]; with regard to problems that it was hard to control the shooting to target area considering the complex sea conditions existing in fire-fighting training of rescue ships, Yang Kenan et al. could realize the

P. Liu · M. Duan · S. Yuan · Y. Dai · Y. Nian · W. Liu (✉)
Institute of Agriculture and Food Standardization, China National Institute of Standardization, Beijing 100191, China
e-mail: liuwen@cnis.ac.cn

S. Ren
School of Computer and Information Technology, Beijing Jiaotong University, Beijing 100044, China

S. Patnaik et al. (eds.), *3D Imaging—Multidimensional Signal Processing and Deep Learning*, Smart Innovation, Systems and Technologies 348,
https://doi.org/10.1007/978-981-99-1145-5_16

3D display technology and human–computer interaction of rescue ships on external fire-fighting training simulation system with the virtual reality equipment, obtaining better virtual simulation and training effects [3]; Zhu Lei et al. had applied the virtual reality technology to assist first-aid operation practical teaching of seriously fractured patients on warships under marine war conditions, solved the "Scene" difficulties for in-situ teaching, explored the teaching resolution under the support of information technology, and strengthened training effects [4]; Yan Yang et al. had applied the virtual reality technology to calibration actual operation training of electric tester calibration device, enabled trainees to have both visual perception and real tactile perception, which greatly improved the sense of reality and training effects of actual operation [5].

In summary, a sort of conventional breakthrough, intuitive, fast, effective, and brand-new training pattern is almost certain. The online training and answering system realizes the remote learning, answering, teaching, and sharing of electronic documents via Internet, ensuring that the trainer and trainees will form a sort of interaction of learning and question–answering on the Internet. Free from temporal and spatial restrictions, this method has the advantages without parallel in traditional training, bringing new opportunities for the development of foods enterprises.

16.2 Main Technologies of Foods Safety Standards Comprehensive Platform

16.2.1 3D Virtual Simulation Technology

WebGL is a sort of JavaScriptAPI [6] that renders the 3D figures in the web browser. Compared with the traditional Web 3D technology, it has two advantages: Firstly, it realizes the Internet interactive 3D animation production via JavaScript script language, without dependence on any browser plug-in; secondly, WebGL is realized based on underlying OpenGL interface and has the function of accelerating the underlying graphics processing unit (GPU) [7].

The process of drawing 3D model with WebGL is as shown in Fig. 16.1.

If directly using WebGL for programming, the development efficiency is lower. At present, there are many WebGL-based 3D graphics application libraries such as Scene.js, Babylon.js, three.js, and Thing.js. Among them, Thing.js is a higher level of IoT 3D library, common methods in WebGL are sealed, and its use is very simple. This paper uses Thing.js as a construction tool of 3D scene. Common functions of Thing.js frame are as shown in Table 16.1.

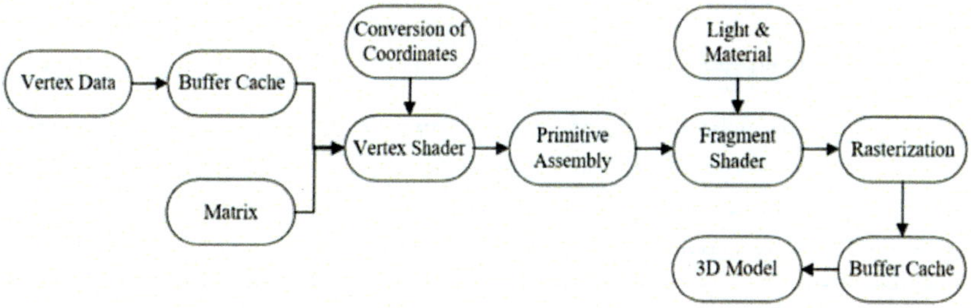

Fig. 16.1 Process of drawing 3D model with WebGL

Table 16.1 Frame functions of Thing.js

API	Content
Namespace	Avoid dispute of function names in different libraries
Core object	As the entry object of App Thing. js library, provide the loading scene and camera control, etc.
Scene level	Scene level management

16.2.2 Text Mining Technology

The text mining technology is such a technology that obtains the useful information [8] from the non-structural text with a certain technology. The text mining technology involves multi-disciplinary fields, including technologies such as text information processing and analysis, text pattern and type identification and distinguishing, machine learning and data mining [9, 10].

For research on text mining, we will extract the text information content properties [11–13] via the text pre-processing, classification, clustering, and visualization. We need to extract the fixed quantity of keywords from each text, and paragraphs that can best reflect the content of text [14]. The term frequency–inverse document frequency (TF-IDF algorithm) is a sort of statistics-based computing algorithm.

The common computing formula of TF is as shown in Formula (16.1), where the frequency of term i in document j is represented by n_{ij}. Generally, the normalized method will be used to process the term frequency, ensuring the stability of computing results.

$$tf_{ij} = \frac{n_{ij}}{\sum_k n_{ij}} \tag{16.1}$$

The common computing formula of IDF is as shown in Formula (16.2). The quantity of documents of term i appearing in document set is represented by $|D|$. In order to avoid the denominator of 0 resulted from the language database excluding the new term, the Laplace Smoothing method is adopted to process the denominator by adding 1.

$$idf_i = \log\left(\frac{|D|}{1+|D_i|}\right) \tag{16.2}$$

TF-IDF algorithm is the combination of TF algorithm and IDE algorithm. The specific computing formula is as shown in Formula (16.3).

$$tf_{ij} \times idf(i, j) = tf_{ij} \times idf_i = \frac{n_{ij}}{\sum_k n_{ij}} \times \log\left(\frac{|D|}{1+|D_i|}\right) \tag{16.3}$$

16.2.3 Knowledge Mapping Technology

Similar to drawing, knowledge mapping is the knowledge network consisting of many nodes and edges and has rich semantic information. In the knowledge network, the connection relationship of nodes is represented by edges, and entities will be represented by nodes. Knowledge mapping construction method includes the top-down method and bottom-up method [15, 16].

The construction process of knowledge mapping includes following aspects, such as knowledge extraction, knowledge integration, inference, knowledge storage and actual application of knowledge mapping [17]. The knowledge acquisition is the prerequisite of constructing the knowledge mapping. The research methods of named entity recognition include rules and dictionaries based method, statistics-based method, and depth learning-based method [18]. The statistics-based method includes the support vector machine, hidden Markov model, and conditional random fields [19]. The recognition accuracy rate of location name, human name, organizational name, and professional field name is higher by utilizing the combined method of Part-Of-Speech tagging and named entity recognition [20]. The relational extraction method falls into following types: formwork-based method, supervised learning-based method, and weak supervised learning-based method [21]. The knowledge mapping storage method has three forms, namely relational database, RDF triplet database, and graph database [19]. The relational database is less flexible than non-relational data storage; the future repair of RDF database is not convenient; the traversal efficiency is high enough and supports the management of affairs [22]. At the knowledge integration link [23], it is required to integrate the differential data, supplement the knowledge mapping, and update and remove duplicate for some wrong information.

16.3 Design of Foods Safety Standards Comprehensive Platform System

As shown in Fig. 16.2, the system architecture integrates the architecture idea of advanced software system, with characteristics such as multi-layered distributed application model, software-based components, unified complete model, and flexible affairs handling and control.

Develop the "Intelligent Scene-specific Foods Safety Standards Training and Implementation Evaluation Comprehensive Platform," including the foods safety standards human–machine interaction question–answering system, intelligent scene-specific foods safety standards training and implementation evaluation platform of foods safety supervisors, intelligent scene-specific foods safety standards training and implementation evaluation platform of foods practicing supervisors, and information-based management module of foods safety standards knowledge library, and integrate the above system platforms into the "Intelligent Scene-specific Foods Safety Standards Training and Implementation Evaluation Comprehensive Platform System." This system includes seven submodules such as homepage module, online training module, human–machine interaction question–answering module, scene simulation module, online communication module, online evaluation module, and backstage management function module.

The system video playback response time is less than 3 s, text interaction time is less than 2.5 s, and data query response time is less than 2 s.

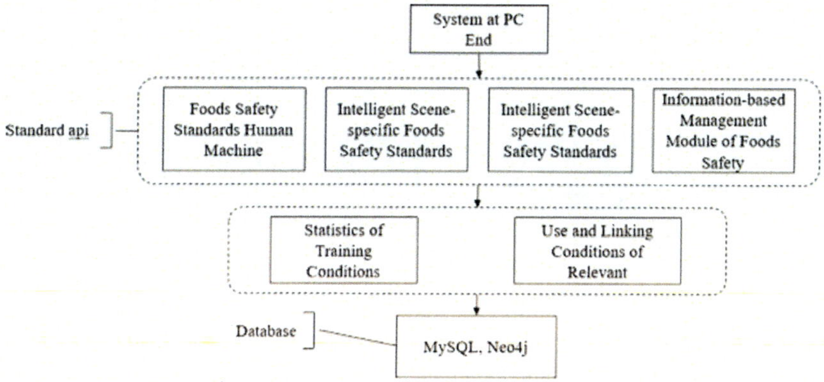

Fig. 16.2 System architecture

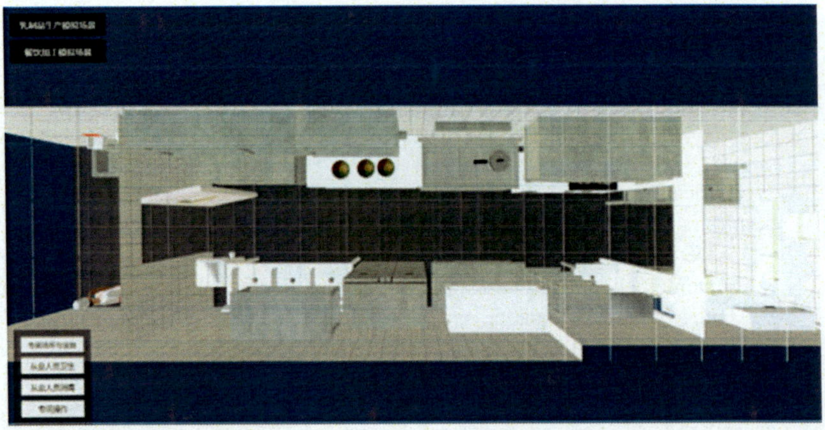

Fig. 16.3 Scene Q&A module

16.4 Functions of Foods Safety Standards Comprehensive Platform System

16.4.1 Foods Safety Standards Human Machine Interaction Question–Answering Subsystem

Text Q&A. It supports the user to enter the question, and the system gives the text answer. If the system doesn't answer the question, such question will be refreshed by the system as new question.

Scene Q&A. It supports the interaction between virtual simulation system and user, enables the foods supervision users to watch the simulation scene of dairy products production and foods processing in an online manner, and also supports the interaction between users and production simulation scene. It is possible to jump into the corresponding scene for explanation by clicking the navigation on the lower left corner and lower right corner (Fig. 16.3).

16.4.2 Intelligent Scene-Specific Foods Safety Standards Training and Implementation Evaluation Subsystem of Foods Safety Supervisors

Data obtaining. Provide the search function of text data and video data of foods supervision users in the manner of catalogue. Search the required data by entering keywords in the text box, and locally download the data by clicking the download button on the right side of the data.

Online training. Provide the function of enabling foods supervision user to obtain text data and video data, browse the selected text data, and play the selected video data. The user may suspend the play and carry out the training learning of expanded knowledge for human interaction function. Make the search by entering the keywords of the required data in the text box, and screen the video data or text data via the check box below the text box.

Online communication. Enable the foods supervision user to select the module, submit new topics, and reply existing topics under its own permission. Release by entering its own topics in the input box, delete topics via the deletion button under its own topics, and express comments for topics via the "Comment" button.

Online evaluation. Enable the foods supervision user to carry out the online answering and evaluation of questionnaire questions, submit the online questionnaire after answering the questions of evaluation questionnaire, and also support the anonymous submission of questionnaire.

Scene simulation. Enable the foods supervision users to watch the simulation scene of dairy products production and foods processing in an online manner, and also support the interaction between users and production simulation scene. It is possible to jump into the corresponding scene for explanation by clicking the navigation on the lower left corner and lower right corner (Fig. 16.4).

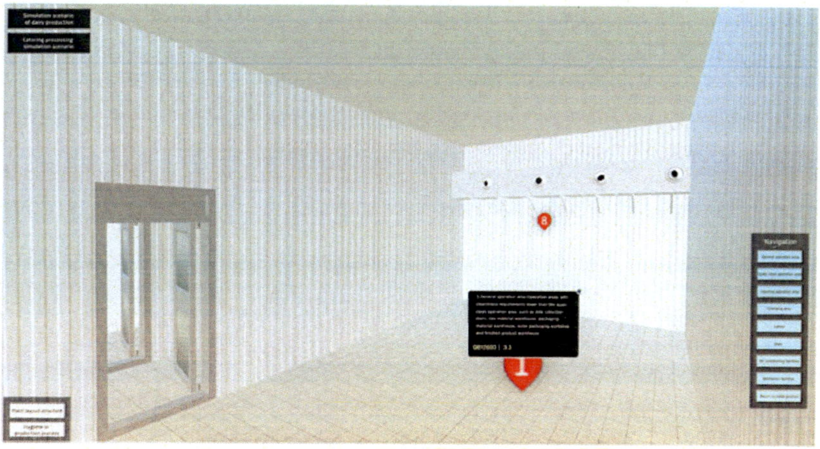

Fig. 16.4 Scene Q&A module

16.4.3 Intelligent Scene-Specific Foods Safety Standards Training and Implementation Evaluation Subsystem of Foods Practitioners

Data obtaining. Provide the search function of text data and video data of foods practicing users in the manner of catalogue. Search the required data by entering keywords in the text box, and locally download the data by clicking the download button on the right side of the data.

Online training. After providing the function of enabling foods practicing user to obtain text data and video data, browse the selected text data, and play the selected video data. The user may suspend the play and carry out the training learning of expanded knowledge for human interaction function. The user may make the search by entering the keywords of the required data in the text box and screen the video data or text data via the check box below the text box.

Online communication. Enable the foods practicing user to select the plate, submit new topics, and reply existing topics under its own permission. Release by entering its own topics in the input box, delete topics via the deletion button under its own topics, and express comments for topics via the "Comment" button.

Online evaluation. Enable the foods practicing user to carry out the online answering and evaluation of questionnaire questions, submit the online questionnaire after answering the questions of evaluation questionnaire, and also support the anonymous submission of questionnaire.

Scene simulation. Enable the foods practicing users to watch the simulation scene of dairy products production and foods processing in an online manner and also support the interaction between users and production simulation scene. It is possible to jump into the corresponding scene for explanation by clicking the navigation on the lower left corner and lower right corner.

16.4.4 Foods Safety Standards Knowledge Library Information-Based Management Subsystem

Management of knowledge library. Provide the repair management and search function of foods safety standards knowledge library by system administrator. Realize the human machine interaction knowledge library management. The system administrator answers new problems entered by foods safety answering user and adds them into foods safety standards knowledge library. Realize the management of database. The system administrator may upload, edit, preview, or delete data. Realize the label management. The system administrator may add the label, set the label status, and link the data and problem information for labels.

Knowledge linkage. Provide the knowledge linkage of foods safety standards knowledge library by system administrator and select the linkage between relevant knowledge and designated knowledge. Connect the relevant knowledge with arrows,

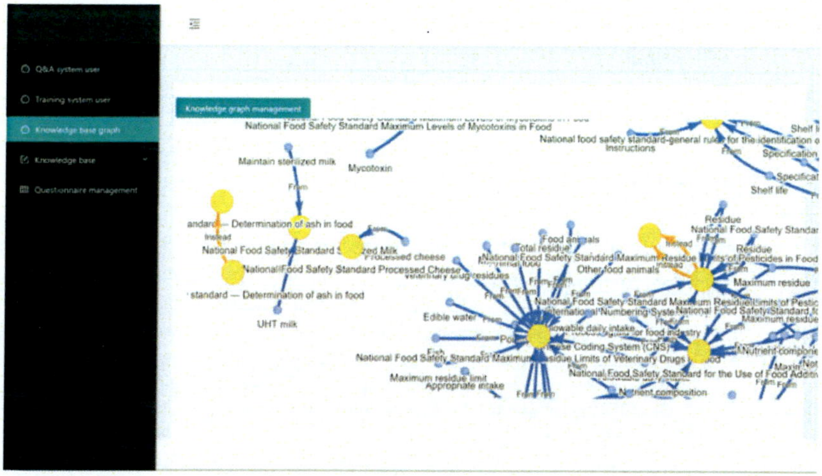

Fig. 16.5 Mapping management of knowledge library

drag them with mouse, and place the mouse on the knowledge node to display relevant information (Fig. 16.5).

Questionnaire management. Provide the design and management of foods safety standards training implementation effects evaluation questionnaire by system administrator. The administrator may look at the completion status of questionnaire, release basic information, and carry out the addition, modification, statistics, and deletion operation for the questionnaire. Click the "Statistics" button behind the item, make statistics for filling status of the questionnaire, and check the detailed information.

16.5 Conclusions

In light of problems that it is hard for workers within the foods safety fields to intuitively and effectively learn and absorb standards knowledge, this paper develops foods safety standards interactive question–answering system and intelligent scene-specific foods safety standards training and implementation evaluation platform, including foods safety standards human–machine interaction question–answering system, intelligent scene-specific foods safety standards training and implementation evaluation platform of foods safety supervisors, intelligent scene-specific foods safety standards training and implementation evaluation platform of foods practitioner supervisors, and information-based management module of foods safety standards knowledge library by using the text mining technology, 3D virtual simulation technology, human–machine interaction technology, and knowledge mapping technology, realizing the diversified training of foods safety standards.

Acknowledgements National Key R&D Program of P. R. China (2019YFC1605201) is gratefully acknowledged.

References

1. Hong, Y., Zhou, K.P., Liang, Z.P., et al.: Development of non-coal mine fire emergency training system based on VR technology. Gold Sci. Technol. **27**(04), 629–636 (2019) (In Chinese)
2. Liu, D.W., Jia, H.R., Jian, Y.H., et al.: Construction and research of emergency training system for tunnel fire based on virtual reality technology. J. Saf. Sci. Technol. **15**(02), 131–137 (2019) (In Chinese)
3. Yang, K.N., Zhang, J.K., Ye, Z.W., et al.: External fire-fighting training simulation system of rescue ships based on virtual reality technology. J. Shanghai Marit. Univ. **40**(02), 95–100+108 (2019) (In Chinese)
4. Zhu, L., Yang, P., Lu, J.J., et al.: Effects observation of application of virtual reality technology to first aid operation training of fractured patient. People's Mil. Surg. **64**(12), 1198–1203 (2021) (In Chinese)
5. Yan, Y., Yang, Y.H., Song, Y., et al.: Application research of virtual reality mapping and visual tactile perception in calibration of virtual electrical equipment tester. Eng. J. Wuhan Univ. **54**(12), 1159–1165 (2021) (In Chinese)
6. Terry, L.T., Steve, Z., Sean, J., et al.: Mobile gyro control for intuitive manipulation of virtual anatomy specimens. FASEB J. **36** (2022)
7. Liu, Z., Gu, X., Dong, Q., Tu, S., Li, S.: 3D visualization of airport pavement quality based on BIM and WebGL integration. J. Transp. Eng., Part B: Pavements **147**(3), 04021024 (2021)
8. Liu, Y.: Research on robust sentiment classification for text-mining-based online shopping product comments. Jiangxi University of Finance and Economics (2021) (In Chinese)
9. Yu, B.H.: Research on key technology for application of semantic internet of things. Chinese Academy of Sciences (Shenyang institute of computing technology, Chinese academy of sciences) (2021) (In Chinese)
10. Meng, L.S.: Research on user's product comments based on text mining technology. Hebei University of Economics and Business (2022) (In Chinese)
11. Hu, Y., Zhang, T.: Land engineering fund project text mining data collection and preprocessing method. Int. Core J. Eng. **8**(6), 42–46 (2022)
12. Xia, C.Y.: Emotional analysis on COVID-19 MicroBlog opinions based on computer learning and depth learning. Jiangxi University of Finance and Economics (2021) (In Chinese)
13. Patel, A., Jain, S.: Present and future of semantic web technologies: a research statement. Int. J. Comput. Appl. 1–10 (2019)
14. Zhang, P.P.: Research on impacts of online public opinions on fluctuations of stock price of listed companies, Donghua University (2020) (In Chinese)
15. Cheng, S.J.: Research on and realization of Thangka characters Q&A system based on knowledge mapping, Northwest Minzu University (2022) (In Chinese)
16. Fan, Q., Tan, G.X., Zhang, W.Y.: The description and application of digital cultural resources based on metadata-a case study of Hubei Digital Cultural Museum. Res. Libr. Sci. **02**, 48–59 (2022). (In Chinese)
17. Tang, L., Guo, C.H., Chen, J.F.: Review of Chinese word segmentation studies. Data Anal. Knowl. Discov. **4**(Z1), 1–17 (2020) (In Chinese)
18. Zhao, S., Luo, R., Cai, Z.P.: Survey of Chinese named entity recognition. J. Front. Comput. Sci. Technol. **16**(02), 296–304 (2022) (In Chinese)
19. Hang, T.T., Feng, J., Lu, J.M.: Knowledge graph construction techniques: taxonomy, survey and future directions. Comput. Sci. **48**(02), 175–189 (2021) (In Chinese)

20. An, Y., Xia, X.Y., Chen, X.L., et al.: Chinese clinical named entity recognition via multi-head self-attention based BiLSTM-CRF. Artif. Intell. Med. **127**, 102282 (2022)
21. Zhu, Y., Ling, Z.G., Zhang, Y.Q.: Research progress and prospect of machine vision technology. J. Graphics **41**(06), 871–890 (2020) (In Chinese)
22. Hong, E.H., Zhang, W.J., Xiao, S.Q., et al.: Survey of entity relationship extraction based on deep learning. J. Softw. **30**(06), 1793–1818 (2019) (In Chinese)
23. Zhu, L.Y., Zhang, J., Hong, L., et al.: Knowledge graph in the field of digital humanities: research progress and future trends. Knowl. Manage. Forum **7**(01), 87–100 (2022) (In Chinese)

Chapter 17
Online Fault Diagnosis of Chemical Processes Based on Attention-Enhanced Encoder–Decoder Network

Qilei Xia, Haiou Shan, Lin Luo, and Zhenhua Zuo

Abstract The data of chemical processes often contain dynamic timing characteristics, and traditional fault detection has low usage of dynamic information, which limits the fault diagnosis performance. To address this problem, this paper proposes a new chemical process fault diagnosis method based on an attention-enhanced encoder–decoder network model (AEDN). The long short-term memory (LSTM)-based encoding part is used to extract the feature information of the process data and combine it with the attention mechanism to utilize the dynamic information among the process data more effectively, the decoding part uses the LSTM and combines the context vector provided by the attention mechanism to provide more accurate state information for the softmax regression, and finally, the softmax regression is used to obtain the probability value of the fault category for each sample data. The attention mechanism improves the model's efficiency in processing dynamic information in the time domain. The proposed method is experimented using Tennessee Eastman (TE) process data and compared with the standard vanilla long short-term memory with batch normalization (LSTM-BN) results. The results show that the proposed method is more effective in diagnosing faults.

17.1 Introduction

Due to substantial technical breakthroughs, modern industrial processes are moving in the direction of enormous scale and complexity. Industrial processes have significant nonlinear and strong coupling characteristics; however, the volume of process data is getting larger and more complicated [1, 2]. It is essential to keep an eye on

Q. Xia (✉) · H. Shan · L. Luo
School of Information and Control Engineering, Liaoning Petrochemical University, Fushun 113001, Liaoning, People's Republic of China
e-mail: 1490145635@qq.com

Z. Zuo
Olefin Plant of Fushun Petrochemical Company, China National Petroleum Corporation, Fushun 113001, Liaoning, People's Republic of China

S. Patnaik et al. (eds.), *3D Imaging—Multidimensional Signal Processing and Deep Learning*, Smart Innovation, Systems and Technologies 348,
https://doi.org/10.1007/978-981-99-1145-5_17

the production process in order to guarantee the machinery is operated safely and to increase production efficiency and quality [3].

Recently, there has been a breakthrough in feature extraction and state recognition thanks to the application of machine learning in scientific and technological research, industrial production, and other domains. The process industry has used deep learning for defect diagnostics. Neural networks perform better at modeling complicated relationships in industrial processes than shallow models do [4–6]. A deep convolutional neural network with deconvolution and a deep autoencoder (DDD) was designed by Kanno and Kaneko [7] and its efficacy in defect diagnosis was confirmed by experiments on the TE process dataset. In order to discover and isolate inferential faults, Luo et al. [8] designed a Bayesian method utilizing the spike-and-slab regularization method. Long short-term memory (LSTM) and ladder autoencoder (LAE) were combined by Zhang and Qiu [9], who creatively proposed LSTM-LAE. When used on the benchmark Tennessee Eastman process, LSTM-LAE demonstrated cutting-edge fault diagnosis performance and accurately localized faults to their relevant variables.

Although the LSTM performs well in solving the long-term dependence problem of sequences, it gradually becomes less effective as the input sequence grows. This is due to the poor design results of the original model, where all the contextual information is limited, and the capability of the whole model is similarly limited. In order to make the model more adequate to make more accurate judgments, the model efficiently maintains the association between the operating conditions and the local temporal properties of the signal by including the attention mechanism. The attention value filters the sensitive information on the hidden layer, and the state vector at each position has to be computed with all other state vectors for attention weights respectively so that the state vector at each position has the global semantic information. Therefore, adding additional layers that can use dynamic information is necessary, which is a different way to ensure the secure operation of chemical systems.

In light of these, this study suggests a deep network with an attention mechanism to improve fault recognition precision. The coding part and decoding part use LSTM networks to extract the feature information of fault condition data and normal condition data through the coding part with the standardized training data and test data, and the attention mechanism introduced as an additional layer makes more effective use of the dynamic information of the local process, while the decoding part can provide more accurate state information for softmax regression, and finally, the probability value of the fault category for each sample data is obtained by using softmax regression. The inclusion of the attention mechanism allows the encoder–decoder network to more effectively filter out the information that is important to the impact of the results from the complex long intervals of information, using contextual vectors to focus the model arithmetic on this more important information. To verify the effectiveness of the proposed method, the proposed model was compared with vanilla LSTM-BN using the Tennessee Eastman chemical process dataset, and the experimental results show that the proposed AEDN model has a higher diagnostic accuracy for each fault mode.

17.2 LSTM Network

A network that places a high value on sequence information is the recurrent neural network (RNN), which is well suited for processing time series data. The conventional recurrent neural network, however, has a flaw that cannot be overlooked. RNN is prone to the problem of gradient disappearance and is prone to forgetting the fault information quickly. As an RNN variation, the LSTM network partially mitigates the issue of gradient explosion or disappearance.

An internal state $\mathrm{m_t} \in \mathbb{R}^D$ in LSTM is dedicated to message-passing updates. The nonlinear output is passed to the external state $S_t \in \mathbb{R}^D$ of the hidden layer. LSTM introduces gating mechanisms to control the path of information transfer: input gates, forgetting gates, and output gates, respectively. It is given in the following form,

$$
\begin{aligned}
f_t &= \sigma\left(\mathbf{W}_{sf}\mathrm{S}_{t-1} + \mathbf{W}_{xf}\mathrm{x}_t + \mathrm{b}_f\right)\\
\mathrm{i}_t &= \sigma\left(\mathbf{W}_{si}\mathrm{S}_{t-1} + \mathbf{W}_{xi}\mathrm{x}_{t-1} + \mathrm{b}_i\right)\\
o_t &= \sigma\left(\mathbf{W}_{so}\mathrm{S}_{t-1} + \mathbf{W}_{xo}\mathrm{x}_t + \mathrm{b}_o\right)
\end{aligned}
$$

where $f_t \in [0, 1]^D$, $\mathrm{i}_t \in [0, 1]^D$, $o_t \in [0, 1]^D$ are the forget gate, the input gate, and the output gate, respectively. $\odot$ is the elementwise multiplication. $\sigma(\cdot)$ is the ReLU function. $\mathbf{W}_{si}$, $\mathbf{W}_{xi}$ and b_i are the weight matrix and additive bias terms of the input gate, respectively. $\{\mathbf{W}_{sf}, \mathbf{W}_{xf}, \mathrm{b}_f\}$ and $\{\mathbf{W}_{so}, \mathbf{W}_{xo}, \mathrm{b}_o\}$ are also the weight matrix and additive bias terms of the forget gate and the output gate, respectively.

$$
\begin{aligned}
\tilde{\mathbf{m}}_t &= \tanh\left(\mathbf{W}_{sm}\mathrm{S}_{t-1} + \mathbf{W}_{xm}\mathrm{x}_t + \mathrm{b}_m\right)\\
m_t &= f_t \odot m_{t-1} + i_t \odot \tilde{m}_t\\
\mathrm{S}_t &= o_t \odot \tanh\left(\mathrm{m}_t\right)
\end{aligned}
$$

where $\tilde{\mathrm{m}}_t \in \mathbb{R}^D$ is the candidate state obtained by nonlinear function. $\mathbf{W}_{sm}$, $\mathbf{W}_{xm}$, and b_m are the weight matrix and additive bias terms of candidate state. m_t stands for internal memory unit. The input gate i_t indicates how much information of the candidate state $\tilde{\mathrm{m}}_t$ that controls the current time needs to be saved. The input gate indicates how much information needs to be saved to control the candidate state $\tilde{\mathrm{m}}_t$ at the current moment. The forget gate f_t indicates how much information needs to be forgotten to control the internal state m_{t-1} at the last moment. The forgetting gate determines which information is forgotten or continued to be remembered by the memory unit. The output gate o_t is used to regulate the amount of information from the internal state m_t that has to be output to the external state S_t at the moment. The output gate can determine which information can be output. For any time t, the internal state of LSTM network records the data information up to the current time.

17.3 AEDN Method for Sequential Fault Diagnosis

The structure of the model is shown in Fig. 17.1. Firstly, the encoder part composed of LSTM network is used to extract the temporal features in the process data to obtain the feature vector $\mathbf{S}'$. The LSTM network uses the back-propagation algorithm to update each parameter value. $a_{t,j}$ indicates how much attention should be paid to the information at j when processing the data at t. $a_{t,j}$ is calculated as follows:

$$a_{t,j} = \frac{\exp(f(\mathrm{S}', \mathrm{S}))}{\sum_{j=1}^{K} \exp(f(\mathrm{S}', \mathrm{S}))}$$

$$f(\mathrm{S}', \mathrm{S}) = \tanh(\mathbf{W}\mathrm{S}_{t-1} + \mathbf{U}\mathrm{S}_j^{'T} + b)$$

where $f(\mathrm{S}', \mathrm{S})$ s is obtained by training a fully connected network, **W** and **U** are learnable parameters. The hidden state S is obtained from the decoding part, and the hidden state S′ is obtained from the encoding part. $f(\mathrm{S}', \mathrm{S})$, also called scoring

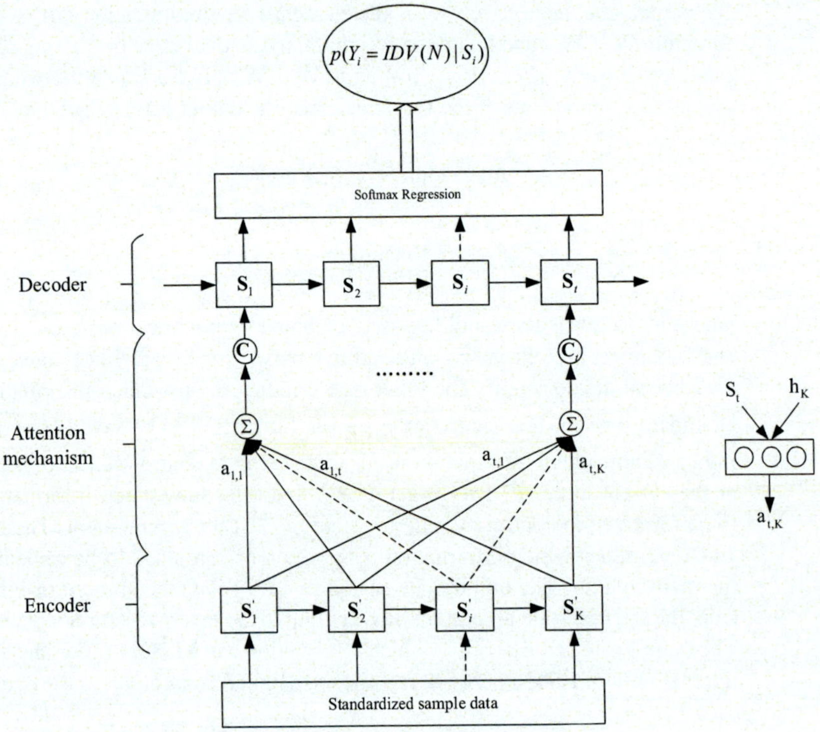

Fig. 17.1 Structure of AEDN model

function, is used to calculate the similarity between the state vector $S^{'}$ and the feature vector S. After we obtaining $f(S^{'}, S)$, use softmatx function to normalize it, convert the values of these weights $a_{t,j}$ to a probability distribution with a range of [0, 1]. Then use the following formula to calculate the attention value C_t. The calculation method is as follows:

$$C_t = \sum_{j=1}^{K} a_{t,j} S_j^{'}$$

When processing the information at t, the decoding part will use C_t and the previous moment memory cell information as input to the LSTM to jointly generate a new hidden state S_i. Finally, softmax regression is used to estimate the probability value of each possible fault class. The calculation is as follows:

$$p(Y_i = N|S_i; \boldsymbol{\theta}) = \frac{e^{\theta_j^T S_i}}{\sum_{l=1}^{N} e^{\theta_j^T S_i}}$$

where $\theta_1, \theta_2, \ldots, \theta_i \in \mathbb{R}^{t+1}$ is the parameter of the model.

17.4 Case Study on Benchmark Process

17.4.1 TE Process Dataset

The Tennessee Eastman (TE) process is a chemical process simulator. It can generate a large amount of normal data while also being able to obtain multiple failure data by setting up multiple failure scenarios. As a public data source, The five basic components of the process are a flash separator, a recycle compressor, a condenser, an exothermic two-phase reactor, and a reboiler stripper. The TE process involves four main gaseous materials in the reaction, yielding two products, a by-product, and an inert gas. There are totally 52 measurement variables, including 41 measured variables and 11 control variables. The TE process defines 21 different types of faults (IDV1-IDV21).

In the case of failure operation, a test dataset can be generated for a failure mode. Each test dataset contained 52 observed variables and 960 observations, of which the first 160 samples were under normal operating conditions and the 161st to 960th samples were the fault data. The sampling interval was 3 min. For monitoring, every variable was used. The training and test samples' values were adjusted to have a zero mean and a single unit of variance.

Table 17.1 Potential parameters and structures for the two methods

	Vanilla LSTM-BN	AEDN
Architecture	{2, 10, 32, 10, 2}	{32, 64, 128, 64, 32}
Optimizer	Adam	Adam
Learning rate	0.01	0.01
Number of epochs	110	110
Batch size	50	50

17.4.2 Diagnostic Results and Discussion

In order to evaluate the performance of vanilla LSTM-BN and the AEDN, we apply these two models to the TE chemical process dataset separately. The TE dataset is more applied in fault diagnosis nowadays. The whole TE dataset consists of a training set and a test set. The candidate structures and parameters of the AEDN model and the vanilla LSTM-BN model are given in Table 17.1. Regarding the effectiveness of the suggested method's defect detection, we use the fault detection rate (FDR) and false alarm rate (FAR) to determine the diagnostic performance of the model. The specific calculation formula is as follows:

$$\text{FDR} = \frac{\text{Total number of samples with faults detected by the method}}{\text{Total number of samples with faults}} \times 100\%$$

$$\text{FAR} = \frac{\text{Total number of normal samples misreported as faults}}{\text{Total number of normal samples}} \times 100\%$$

The above two methods were used to conduct 10 fault diagnosis experiments on the test dataset. For each test, dataset contains 960 data points, and faults are introduced starting from the 161st data point. After the experiment, the defect detection rate (FDR) and false alarm rate (FAR) were recorded in Table 17.2. As can be seen from Table 17.2, the average fault detection rate of the AEDN model is as high as 94.06%, and the average false alarm rate is 0.249% lower than the other model.

This experiment initially looks at the normalized confusion matrix of the vanilla LSTM-BN model and the AEDN model with regard to the fault detection performance of the suggested method. As shown in Fig. 17.2, the confusion matrix is a specific matrix that can be used to present a visualization of algorithm performance, usually used in supervised learning, in which each row corresponds to the actual class of the target, each column corresponds to the estimated class produced by the

Table 17.2 Evaluation metrics for AEDN and vanilla LSTM-BN

Model	Average FDR/%	Average FAR%
Vanilla LSTM-BN	86.91	0.539
AEDN	94.06	0.249

classifier, and the diagonal elements indicate the percentage of various failure modes that the classifier can correctly classify. The AEDN model has the lowest amount of misclassification probabilities, indicating that it can classify the majority of faults appropriately.

In this experiment, time steps = 50 is used, and the same fault data is input in the same experimental environment for experiment; the ROC curve and PR curve of vanilla LSTM-BN model and AEDN model are obtained. Draw curves with false positive rate (FPR) and true positive rate (TPR) as abscissa and ordinate to obtain ROC curves under different fault types. However, because the ROC curve does not depend on the specific distribution of categories, when the category imbalance occurs, the excessive number of negative cases will make the ROC curve show an overly optimistic effect estimate, so it needs to be further evaluated with the PR curve. Draw PR curve with recall and precision as horizontal and vertical axis.

On Fig. 17.3a–b, are shown the ROC curves of vanilla LSTM-BN and the proposed model, respectively. Inspecting the micro-average AUC-ROC shows the value of the proposed model is the largest value ($AUC_{AEDN} = 0.99 > AUC_{\text{vanilla LSTM-BN}} = 0.96$). For various fault types, the suggested model's ROC curves are closest to the coordinate axis' upper left corner. The diagnostic performance of the suggested model, however, is noticeably superior to the other model for the macro-average AUC-ROC ($AUC_{AEDN} = 0.99 > AUC_{\text{vanilla LSTM-BN}} = 0.95$). At the same time, for fault types that are difficult to diagnose, such as fault 3 ($AUC_{AEDN} = 0.93$), fault 9 ($AUC_{AEDN} = 0.91$), fault 15 ($AUC_{AEDN} = 0.92$), and fault 21 ($AUC_{AEDN} = 0.95$), the AUC values calculated by the proposed model are higher than that of vanilla LSTM-BN. As shown in Fig. 17.4a–b are the P-R curves of vanilla LSTM-BN and the proposed model, respectively. The random performance level of a classifier is represented by the horizontal red lines in Fig. 17.4. From the AUC value of the micro-average and macro-average P-R curve, the proposed method demonstrates the best diagnostic performance, for the micro-average P-R curve, we have $AUC_{AEDN} = 0.93 > AUC_{\text{vanilla LSTM-BN}} = 0.78$, and for the macro-average P-R curve, we have $AUC_{AEDN} = 0.87 > AUC_{\text{vanilla LSTM-BN}} = 0.74$. For faults 3, 9, 15, and 21, AUC value of PR curve of AEDN model is still higher than that of vanilla LSTM-BN.

Moreover, the experiment also carried out the standard macro-F1 scores (macro-F1) to express different preferences for precision/recall, so as to obtain the tradeoff between the two, as shown in Fig. 17.5. It is evident that the F1 score of AEDN is greater than vanilla LSTM-BN for a variety of defects, further demonstrating the proposed method's superior diagnostic accuracy.

17.5 Conclusion

In this paper, an attention-enhanced network based on LSTM is proposed. The Tennessee Eastman data are used to test and verify the suggested model in numerous ways, and it is then contrasted with the other fault diagnostic model. The experimental findings demonstrate that the AEDN model can extract the feature information of

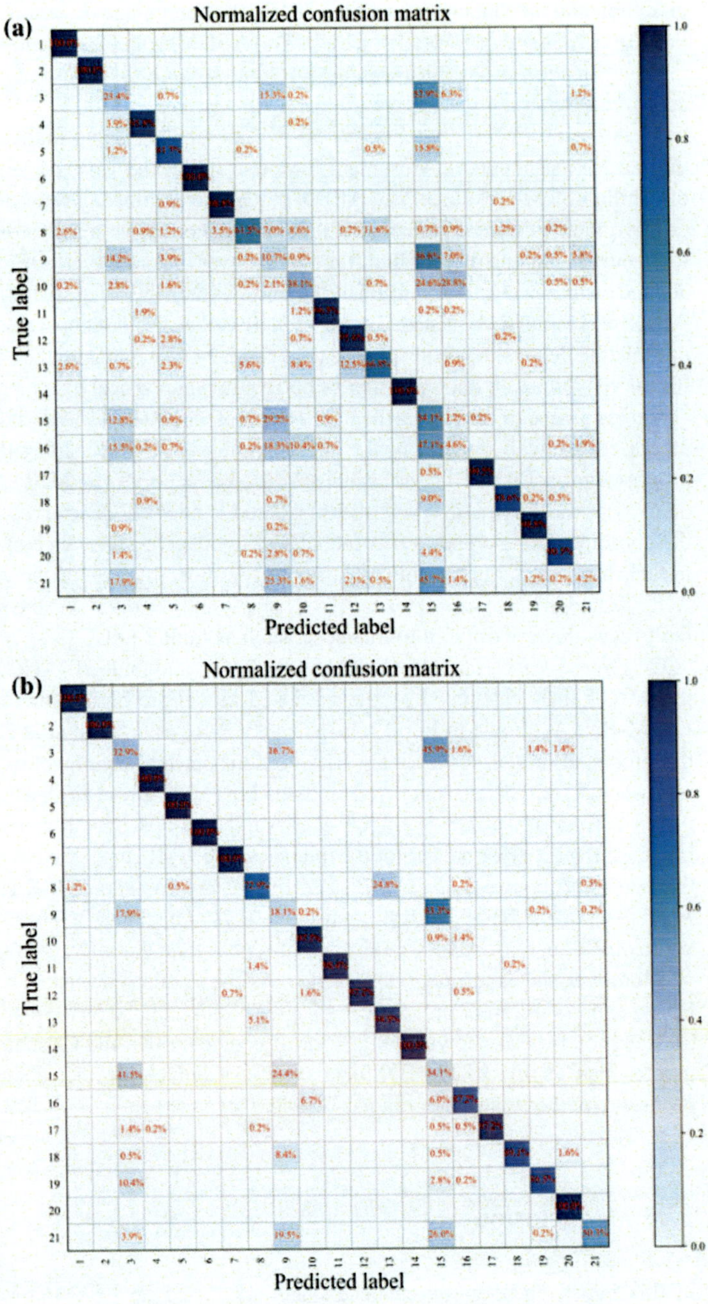

Fig. 17.2 Confusion matrix for two methods: **a** vanilla LSTM-BN and **b** AEDN

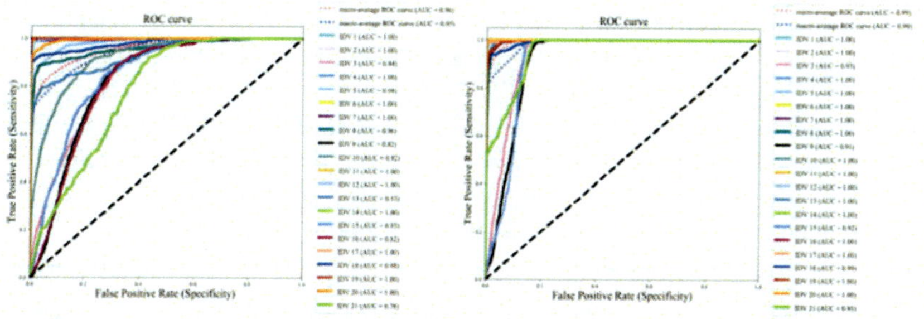

(a) (b)

Fig. 17.3 ROC curves for two methods: **a** vanilla LSTM-BN and **b** AEDN over the different fault modes

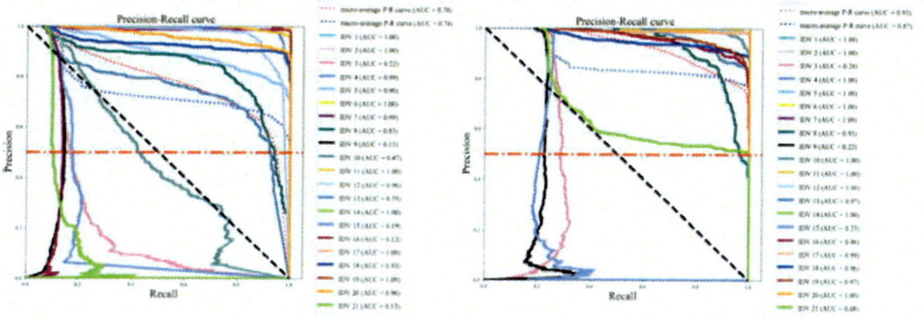

(a) (b)

Fig. 17.4 PR curves for two methods: **a** vanilla LSTM-BN and **b** AEDN over the different fault modes

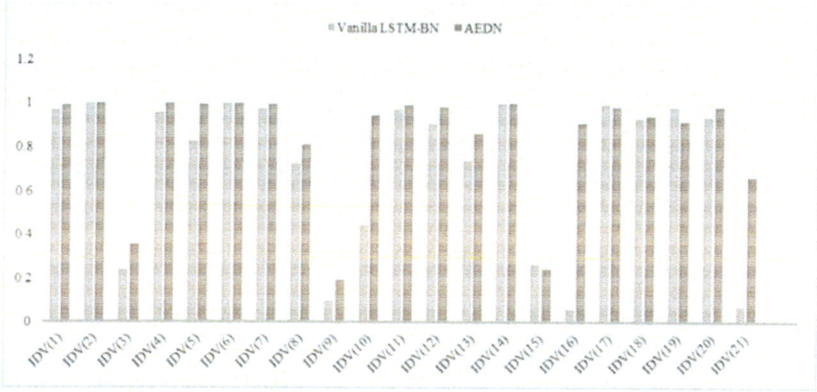

Fig. 17.5 F1 score for two methods

process data using the coding component and does not require human feature extraction. The AEDN model can introduce the feature weight coefficient while extracting the time series feature data by adding the attention mechanism. The decoder weights the features extracted at various scales and produces the fault probability value for each sample by softmax regression, which increases the precision of model fault identification. The AEDN model offers higher nonlinear feature extraction capabilities and a stronger generalization capacity when compared to the other model, which is crucial for fault diagnosis and discrimination. The AEDN model still has opportunity for refinement and optimization, according to experiments. The goal of future research is to strengthen the model's generalization capabilities and further increase the model's diagnostic accuracy.

References

1. Luo, L., Xie, L., Su, H.Y., Mao, F.S.: A probabilistic model with spike-and-slab regularization for inferential fault detection and isolation of industrial processes. J. Taiwan Inst. Chem. Eng. **123**, 68–78 (2021)
2. Luo, L., Xie, L., Su, H.Y.: Deep learning with tensor factorization layers for sequential fault diagnosis and industrial process monitoring. IEEE Access **8**, 105494–105506 (2020)
3. Ge, Z., Song, Z., Gao, F.: Review of recent research on data-based process monitoring. Ind. Eng. Chem. Res. **52**(10), 3543–3562 (2013)
4. Wang, W., Galati, F.A., Szibbo, D.: Gear diagnostics based on LSTM anomaly detection. Int. J. COMADEM **24**(2), 3–13 (2021)
5. Wu, H., Zhao, J.: Deep convolutional neural network model based chemical process fault diagnosis. Comput. Chem. Eng. **115**, 185–197 (2018)
6. Zhou, F., Yang, S., Fujita, H., Chen, D., Wen, C.: Deep learning fault diagnosis method based on global optimization GAN for unbalanced data. Knowl.-Based Syst. **187**, 104837 (2020)
7. Kanno, Y., Kaneko, H.: Deep convolutional neural network with deconvolution and a deep autoencoder for fault detection and diagnosis. ACS Omega **7**(2), 2458–2466 (2022)
8. Luo, L., Xie, L., Su, H.Y.: Process monitoring with sparse Bayesian model for industrial methanol distillation. IFAC-Papers OnLine **53**(2), 424–430 (2020)
9. Zhang, S., Qiu, T.: Semi-supervised LSTM ladder autoencoder for chemical process fault diagnosis and localization. Chem. Eng. Sci. **251**, 117467 (2022)

Chapter 18
Micro-nano Satellite Novel Spatial Temperature Measurement Method and Experimental Study

Dawei Li, Zhijia Li, Zhanping Guo, Zhiming Xu, and Shijia Liu

Abstract To solve the problem of micro-nano satellite about the complexity temperature acquisition network cable connectivity, diversity in temperature acquisition circuits and devices, complicated computation of temperature formula and so on, a single bus thermometry technology based on 1-wire commercial temperature measurement devices DS18B20 instead of the traditional thermometry technology on the satellite was proposed in the paper. In the study, the space measurement system based on DS18B20 including temperature circuit, software and cable network applied to the temperature measurement of the micro-nano satellite was designed, and space vacuum test system was built to study and verify the performance about DS18B20. The results show the devices DS18B20 have a stable performance and accurate temperature measurement. So the commercial temperature measurement devices DS18B20 can meet the temperature measurement requirement of equipment on the satellite. The research in this paper provides a new method for temperature measurement of the equipment on the satellite and lays the foundation for the application of DS18B20 temperature measurement devices on the satellite.

18.1 Introduction

With the development of microelectronic technology [1], the development of micro-nano technology represented by micro-electromechanical system and micro-opto-electromechanical system especially makes it possible to realize the micro satellite, nano satellite and pico satellite [2].

In recent years, with the increasing complexity of functions on small satellites, thermal control systems of small satellites or micro satellites also have an increasing demand for the number of temperature measurements [3]. At present, some small

D. Li · Z. Li · Z. Guo
Beijing Institute of Spacecraft Environment Engineering, Beijing 10094, China

Z. Xu (✉) · S. Liu
DFH Satellite Co., Ltd., Beijing 100094, China
e-mail: xzhm810604@163.com

S. Patnaik et al. (eds.), *3D Imaging—Multidimensional Signal Processing and Deep Learning*, Smart Innovation, Systems and Technologies 348,
https://doi.org/10.1007/978-981-99-1145-5_18

satellites have the demand of more than 200 temperature measurement channels [4]. Using thermistor for temperature measurement, each thermistor needs to lead two temperature measurement cables from the satellite host. Due to the large number of equipments on the satellite, two temperature measurement cables are required from the satellite host to the equipment [5]. The high density characteristics of small satellites, especially micro-nano satellites, makes the point-to-point cable connection complex and difficult to use. The temperature measuring cable network is connected by pin-insertion and jack. This method is easy to cause short circuit and abnormal temperature measurement [6].

According to the above problem, a DS18B20 temperature measurement technology based on 1-wire is introduced in this paper [7]. All the DS18B20 temperature sensors are serially connected by a bus. The value of the temperature sensors are directly read by the software [8]. Although some research work has been carried out in the literature [9], it was only an experimental research and it had not been used in micro-nano satellite. On the above basis, the DS18B20 was applied to measure temperature in the micro-nano satellite in the article. Not only were hard and soft design carried out, but also the design of the cable network was conducted according to the requirement of the real satellite temperature measurement. Experimental study was developed according to the real state of the satellite in orbit.

18.2 Temperature Measurement Principle on DS18B20

DS18B20 device with networking digital temperature sensor is produced by the DALLAS company. It uses the 1-wire bus interface, allowing the users to easily set up sensor networks to introduce the novel concept for temperature measurement system [10]. The DS18B20 sensor element and the temperature conversion circuit are integrated in an integrated circuit such as a transistor size. Internal structure diagram of the device is shown in Fig. 18.1 [11].

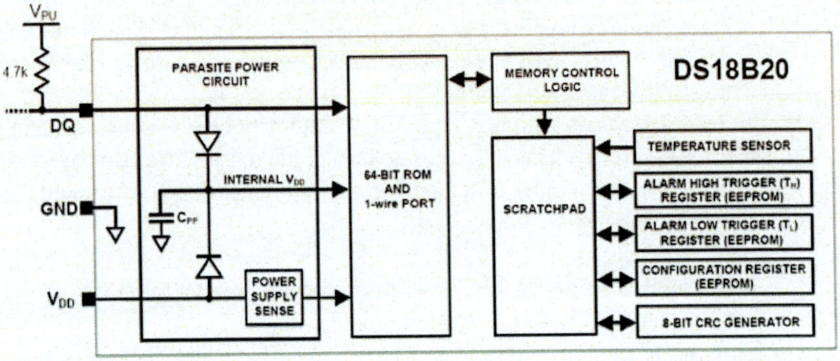

Fig. 18.1 DS18B20 device internal structure block diagram

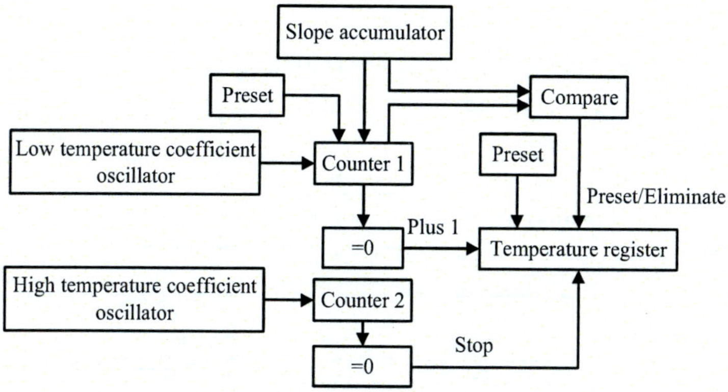

Fig. 18.2 DS18B20 temperature measurement principle block diagram

Temperature measurement principle of the DS18B20 is shown in Fig. 18.2. The oscillator frequency of the low temperature coefficient inside the device in the Fig. 18.2 is little affected by the temperature, and is used to generate a pulse signal with a fixed frequency and send it to the counter 1. The oscillation rate of the high temperature coefficient oscillator changes significantly with temperature variation, which is equivalent to the T/F converter. The generated signal of which is used as the pulse input of counter 2. A base value corresponding to −55 °C is preset in the temperature register and counter 1. Counter 1 subtracts the pulse signal generated by the low temperature coefficient oscillator. When the preset of counter 1 is reduced to 0, the value of temperature register will be increased by 1. The preset of counter 1 will be loaded again, and the counter 1 will start counting the pulse signal generated by the low temperature coefficient oscillator again. This above cycle continues until the counter 2 counts to 0, and the accumulation of the temperature register value is stopped. At this time, the value in the temperature register is the measured temperature value.

18.3 A New Temperature Measurement Experiment of Micro-nano Satellite

18.3.1 Thermoscope System Design on DS18B20

Firstly, the temperature measurement system design was conducted. According to the literature [12], the circuit design and software design were carried out as follows. The temperature measurement system is designed with parasitic power supply and VDD and GND are grounded. When the signal line DQ of 1-wire bus is high, the signal energy is stolen to supply power to DS18B20. The following describes the

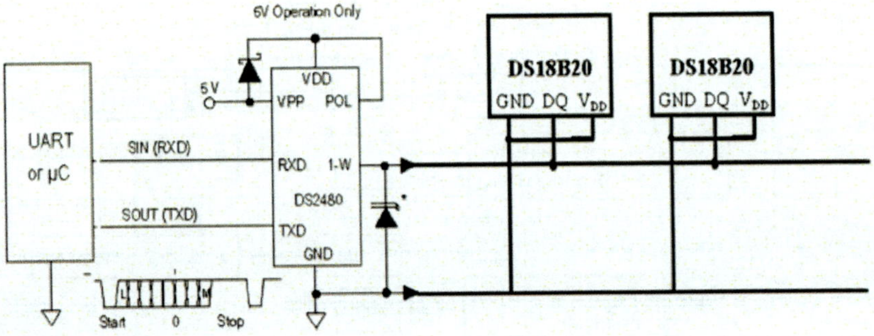

Fig. 18.3 Signal bus temperature acquisition interface circuit block diagram

principle of parasitic power supply mode. The principle is that when the signal line DQ of 1-wire bus is high, the signal energy is stolen to supply power to DS18B20; at the same time, part of the energy is used to charge the internal capacitor. When the DQ is low, the energy is released to supply power to the DS18B20. DS18B20 single bus interface adopts DS2480B interface chip method, which can simplify hardware and software design. The principle is followed as Fig. 18.3.

The software design of the temperature measurement system mainly completes the function of calling and managing the DS18B20, calculating and processing the temperature value and so on. DS18B20 work protocol flow includes initialization, ROM operation instructions and function operation instructions. The working time sequence includes initialization time sequence, write time sequence and read time sequence. Therefore, the master computer must go through three steps to control the DS18B20 to complete the temperature conversion. First, the DS18B20 is reset before each read/write; second, a ROM command is send after successful reset; and finally the function operation command is send.

18.3.2 Design of Temperature Measurement Cable Net

According to the principle of DS18B20 temperature measurement, DS18B20 temperature measurement cable design was conducted. Lead wire specification AF46-200 19X0.12 was adopted. Different branch lengths were designed according to the layout design of satellite equipment. The XW-2 satellite cable network design is shown in Fig. 18.4.

The DS18B20 temperature measurement devices were pasted on the corresponding equipments according to the cable network design showed in Fig. 18.4 and lastly connected to the host server. Meanwhile, thermocouples were attached to the corresponding equipments to measure temperature for comparison with the temperature data of the DS18B20.

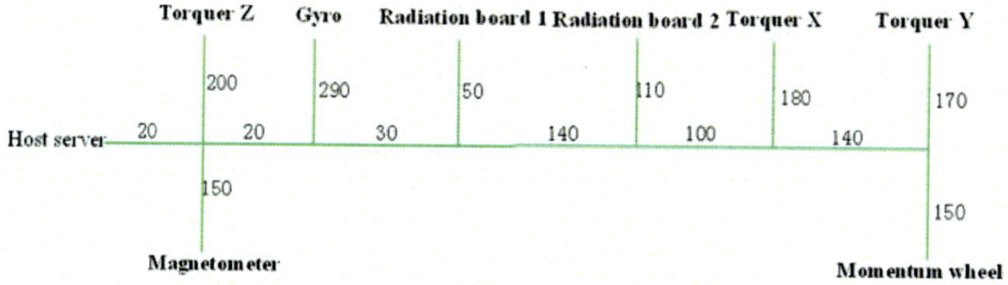

Fig. 18.4 The cable net design-based DS18B20

18.3.3 Temperature Measurement Experiment Based on Micro-nano Satellite

The experiment is mainly to verify the correctness of the thermal control design of micro-nano satellite [13]. A temperature measurement technology based on 1-wire device DS18B20 was used to collect the temperature value of the equipments on the satellite. At the same time, the thermocouple was pasted on the equipment on which the device DS18B20 had been pasted. The temperature measurement accuracy of the new technology is verified by comparing temperature value of the thermocouple with temperature value of the corresponding DS18B20 [14].

The experiment was carried out in a vacuum simulator with a diameter of about 2 m. The vacuum simulator can simulate the space environment; the vacuum degree is better than 1.3×10^{-3} Pa and the temperature of heat sink is better than 100 k. Thermal control implementation of the experiment was conducted according to the thermal control design scheme of micro-nano satellite. All the experiment cables were connected to the corresponding test equipments by the flange of the vacuum simulator and then the vacuum simulator door was closed. The vacuum was started until the vacuum and heat sink temperature can meet the experiment requirement. The experiment was conducted according to the experimental order required. Data acquisition and monitoring were carried out in the experimental process by test data acquisition software of spacecraft. Its sampling period is 32 s. The Schematic of the experimental system was shown in Fig. 18.5.

According to the thermal experiment outline of the XW-2 satellite, two conditions which covered respectively the extreme low temperature and extreme high temperature of the satellite were carried out, and the experiment started from the low temperature condition. The data in the process of the experiment were noted and saved by the data acquisition software.

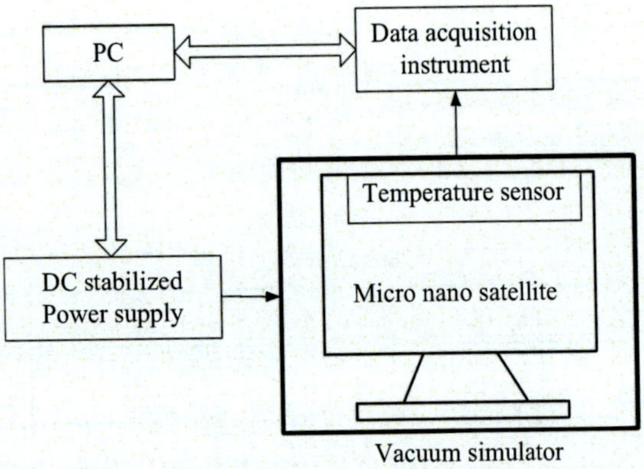

Fig. 18.5 Schematic of the experimental system

18.4 Experimental Results and Analysis

DS18B20 monitoring temperature data in low temperature condition is showed in Fig. 18.6. It can be seen from Fig. 18.6 that the DS18B20s can monitor the temperature of the devices in real time and have good data acquisition continuity. The above results show that the DS18B20 is stable at low temperature.

The data of DS18B20 temperature measuring element at low temperature are showed in Table 18.1 by comparing with the one of thermocouple temperature

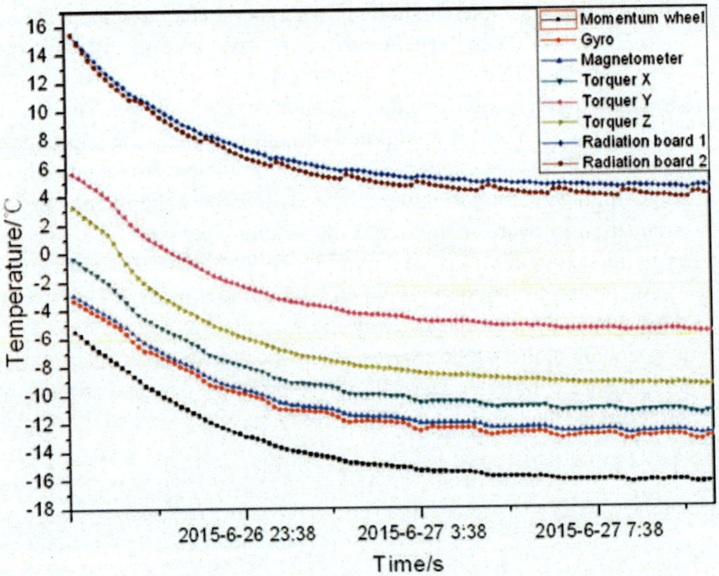

Fig. 18.6 DS18B20 temperature acquisition telemetry data in low temperature condition

measuring element. It can be seen form Table 18.1 that the DS18B20 temperature measuring element has also good temperature measurement performance and the accuracy of temperature measurement at low temperature condition by comparing with the thermocouple usually used in the spacecraft.

DS18B20 monitoring temperature data in high temperature condition is showed in Fig. 18.7. It can be seen from the Fig. 18.7 that the DS18B20s can monitor the temperature of the devices in real time and have good data acquisition continuity. The above results show that the DS18B20 is stable at high temperature.

The data of DS18B20 temperature measuring element are showed in Table 18.2 by comparing with the one of thermocouple temperature measuring element at

Table 18.1 The data comparison between thermocouple and DS18B20 at low temperature (°C)

Equipment	Thermocouple data	DS18B20 data	Thermocouple—DS18B20
Momentum wheel	−14.355	−15.25	0.895
Gyro	−13.235	−13.0625	−0.1725
Magnetometer	−13.388	−12.75	−0.638
Torquer X	−11.208	−11.25	0.042
Torquer Y	−6.355	−5.625	−0.73
Torquer Z	−8.90	−9.0125	0.1125
Radiation board 1	4.416	4.4375	−0.0215
Radiation board 2	4.841	3.9375	0.9035

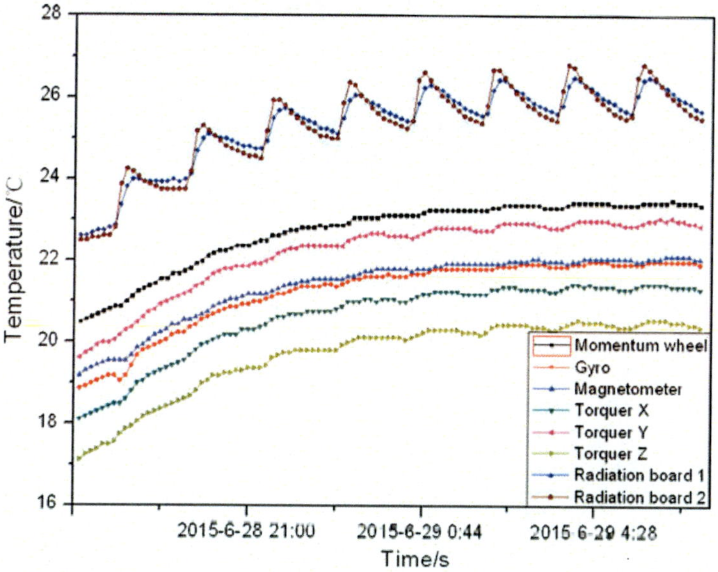

Fig. 18.7 DS18B20 temperature acquisition telemetry data in high temperature condition

Table 18.2 The data comparison between thermocouple and DS18B20 at high temperature (°C)

Equipment	Thermocouple data	DS18B20 data	Thermocouple—DS18B20
Momentum wheel	24.673	23.8125	0.08605
Gyro	22.093	22	0.093
Magnetometer	21.864	22.125	−0.261
Torquer X	21.394	21.4375	−0.0435
Torquer Y	24.12	23.375	0.745
Torquer Z	20.126	20.4375	−0.3115
Radiation board 1	26.016	26	0.016
Radiation board 2	25.833	25.8125	0.0205

high temperature. It can be seen form Table 18.2 that the DS18B20 temperature measuring element has good temperature measurement performance and the accuracy of temperature measurement at high temperature condition by comparing with the thermocouple usually used in the spacecraft.

In the process of micro-nano satellite thermal experiment, the temperature measuring performance and the accuracy of temperature measurement of the DS18B20 were validated based on the low and high condition test in vacuum environment by comparing with the data of thermocouple showed in the Figs. 18.6, 18.7 and Tables 18.1, 18.2. The equipments of temperature range from −55 to 125 °C are suitable for DS18B20 measurement temperature.

18.5 Conclusions

The temperature measurement principle of a new type of space temperature measuring device DS18B20 applied to the micro-nano satellite was studied. The hardware and software design and cable network design were carried out. Lastly, the investigations are validated by the ground experiment. The conclusions can be drawn as follows. Firstly, the temperature measurement consistency between the thermocouple and DS18B20 indicates that the DS18B20 is stable and accurate in the vacuum environment. Secondly, the feasibility of applying the DS18B20 Temperature measurement element to the satellite is verified by the ground experiment. Thirdly, the research of this paper provides a new way to measure the temperature of satellite in orbit.

References

1. Osiander, R., Darrin, M.A.G., et al.: MEMS and microstructures in aerospace applications. CRC Press Taylor & Francis Group, USA (2006)
2. George, T.: Overview of MEMS/NEMS technology development for space applications at NASA/JPL. Proc. SPIE **5116**, 136–148 (2003)
3. Schoiz, A., Ley, W., Dachwald, B., et al.: Flight results of the COMPASS-1 picosatellite mission. Acta Astronaut. **67**, 1289–1298 (2010)
4. Zhang, J.X., Wang, H., Sun, J.L.: The application analysis of thermistor in spacecraft. Chin. Space Sci. Technol. **6**, 54–59 (2004). (In Chinese)
5. Li, Z.W., Dong, J.L., Liu, C., et al.: Development of thin-film thermocouple for measuring transient high temperature on spacecraft surfaces. Spacecraft Environ. Eng. **34**(4), 393–397 (2017). (In Chinese)
6. Li, J.T.: The small satellite housekeeping technology. Chin. Space Sci. Technol. **35**(6), 54–59 (2004). (In Chinese)
7. Zhou, L.Y.: Temperature monitoring system based on DS18B20. Annual Report of China Institute of Atomic Energy, pp. 63–64 (2010)
8. Hou, Z.Q., Hu, J.G.: Spacecraft Thermal Control Technology—Theory and Application. Chinese Science and Technology Press, Beijing (2007). (In Chinese)
9. Min, G.R., Guo, S.: Spacecraft Thermal Control, 2nd edn. Science Press, Beijing (1998). (In Chinese)
10. Gilmore, D.G.: Spacecraft Thermal Control Handbook, 2nd edn. The Aerospace Corporation Press, EI Segundo (2002)
11. Low rate tid test of DS1820 class 1-wire temperature transducers. TEC-EDD/2005.45/GF
12. Xu, Z.M., Liu, P., Chang, X.Y.: Space environment adaptation experiment of commercial sensor Ds18B20. Appl. Mech. Mater. 635–637:760–767 (2014)
13. Li, Z.S., Ma, C.J., Mao, Y.J., et al.: Rapid analysis of temperature filed for orbiting nanosatellites. Spacecraft Environ. Eng. **38**(2), 122–129 (2021) (In Chinese)
14. Li, J.H.: Design a wireless temperature measurement system based on NRF9E5 and DS18B20. In: 2010 International Conference on Measuring Technology and Mechatronics Automation (ICMTMA), pp. 910–913 (2010)

Chapter 19
Research on Plant Allocation of Sponge City Construction Based on Deep Learning

Huishan Wang, Xiang Zhao, and Jiating Chen

Abstract Since the construction of Sponge City in China, the research on the construction of Sponge City has mostly focused on the engineering level, and the research on the selection, function and configuration of plants has lagged behind, resulting in unreasonable plant selection, lack of "localization" of plant design, and lack of systematic and targeted consideration and design of plant landscape configuration, which has restricted the construction and development of Sponge City to some extent. Mian city can reduce urban waterlogging, improve the utilization rate of water resources, and meet the requirements of green environmental protection development in China. Through analysis, the functions of plants in Sponge City are summarized as landscaping and ecological protection. Based on the clues of meeting these two functions, combined with in-depth learning, this paper proposes the principles of plant selection and configuration under the concept of Sponge City, which provides a theoretical basis for the selection of plants in the construction of Sponge City. The paper summarizes the plants that can be used in sponge cities, and draws the common plant configuration patterns in low impact development facilities during the research process, providing relevant theoretical basis and experience for further construction of sponge cities in China.

19.1 Introduction

With the progress of science and technology in modern society, human beings' ability to transform nature has also changed greatly, and gradually lost their awe of nature, and began to exploit, utilize and destroy nature without restriction, resulting in the fragmentation of natural ecological environment, the continuous degradation of ecosystem functions, serious water pollution and increasingly scarce water resources [1]. Biodiversity is also decreasing year by year; With the urbanization, human beings

H. Wang (✉) · X. Zhao · J. Chen
School of Architecture and Civil Engineering, Xiamen Institute of Technology, Xiamen 361021, Fujian, China
e-mail: 790693032@qq.com

S. Patnaik et al. (eds.), *3D Imaging—Multidimensional Signal Processing and Deep Learning*, Smart Innovation, Systems and Technologies 348,
https://doi.org/10.1007/978-981-99-1145-5_19

are constantly migrating to cities, and the urban area is constantly increasing; In order to improve the utilization rate of urban land and continuously compress the area of urban green space, the proportion of hard pavement is increasing [2]. According to the traffic function, status, role of roads in the urban road system, service function to buildings along the road and service frequency of pedestrians, urban roads can be divided into four categories: expressway, trunk road, secondary trunk road and branch road. According to the classification of road activity subjects, roads can be divided into vehicular roads, pedestrian-vehicle mixed roads and pedestrian roads [3]. The road is the foundation of a beautiful and comfortable living environment and urban functions, the link and pulse of urban social and economic activities, and an important window for people to feel the characteristics of urban landscape and display the image of the city. The concept of Sponge City came into being [4].

In October 2014, the Ministry of Housing and Urban Rural Development issued the Technical Guide for Sponge City Construction–Construction of Low Impact Development Rainwater System (for trial implementation) as the technical specification for urban construction and development of sponge cities in China. However, the specific implementation process lacks "localization" and consideration in plant landscape configuration [5]. In view of the uneven distribution of water resources and the shortage of total water resources in China, the rainwater in urban construction is mainly discharged through the municipal pipe network. How to store the rainwater in the "spongy body" in the rainy season and release the water in the "spongy body" for use in the dry season is the core problem in the construction of sponge cities. In this process, the plants that constitute the urban landscape are important materials to form the "sponge" [6]. Different from mechanical passive shallow learning, deep learning focuses on mastering connotation through deep processing, constructing personal knowledge system, effectively migrating to life problems, and promoting the achievement of learning goals and the development of high-level thinking. However, how to effectively organize in-depth learning is a topic of concern for front-line teachers [7].

At present, countries all over the world are paying more and more attention to the construction of Sponge City in the process of actively strengthening modernization. Relevant fields in China have also started to build Sponge City from the perspectives of urban economy, politics, humanities and social development, and some cities have started large-scale construction. From a long-term perspective, all regions in China should continuously combine the current situation of urban development and the concept of Sponge City, comprehensively analyze the climate and urban supporting facilities and other factors, and carry out targeted innovation construction of Sponge City. Therefore, in the process of urban construction and development, we should pay more attention to road greening projects, combine the trend of urban development with the characteristics of the times, design scientifically and pay attention to the aesthetic art in the selection and allocation of road green plants, while enhancing the role of plants in environmental protection.

19.2 Application of Various Plant Landscape Configurations in Sponge Cities

19.2.1 The Role of Plant Landscape in Sponge City

The so-called Sponge City means that in the process of urban construction, attention should be paid to environmental protection, mainly reflected in the collection and management of rainwater, so as to adapt to environmental changes, especially when dealing with disasters caused by environmental changes, it has certain flexibility to reduce losses caused by disasters [8]. As shown in Fig. 19.1.

Flood management is an important part of urban development today. Poor drainage and flood control ability is a common problem in many old cities, especially in rainy season and typhoon weather, which directly affects the safety of people's lives and property [9]. Therefore, in the development of modern cities, it is not only an important factor to ensure the safety of cities but also an important measure to improve the utilization rate of water resources, and it is also an important embodiment of urban environmental protection to strengthen the management of rainwater, to maximize the ability of rainwater collection and management and to solve the problems of water absorption, water storage, water seepage and water purification [10].

Sponge City refers to a city that can well adapt to extreme climate and environmental changes. Like a sponge, it can store and purify natural rainfall for emergencies.

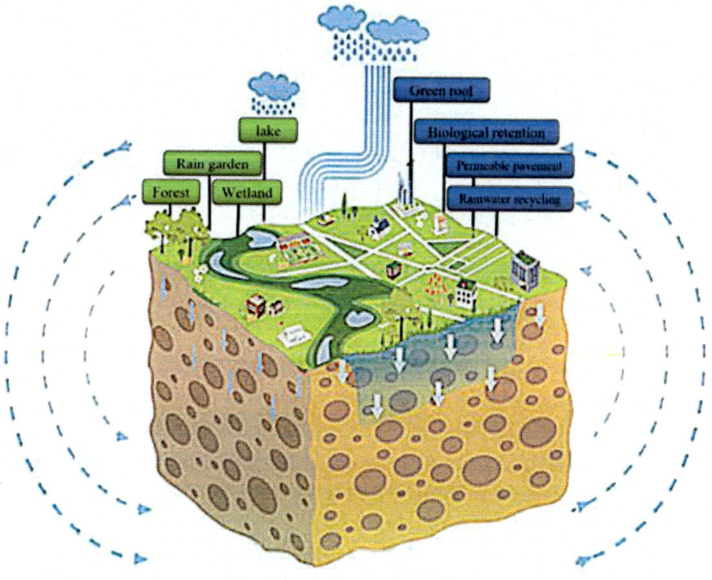

Fig. 19.1 Schematic diagram of Sponge City

The construction of the Sponge City cannot be separated from the natural and harmonious ecological environment. People use various means to transform the city into an "ecological factory" integrating drainage, water storage and water purification, so as to make efficient use of rainwater resources and protect the ecological environment. Under the concept of Sponge City, the following aspects should be done to realize the effective allocation of plants: (1) Planting grass ditch is an important way to realize the construction of Sponge City. It refers to the surface ditch planted with vegetation, which can replace hardened ditch, dredge surface runoff, effectively purify rainwater and improve the quality of landscape environment. (2) The sunken green space refers to the green space lower than the surrounding road surface, such as rain wetlands and infiltration ponds. Sinking greenbelt can collect rainwater, reduce surface runoff, avoid urban waterlogging and save irrigation water for greenbelt. (3) Rainwater garden is an important method to realize the construction of Sponge City. Rainwater garden can be used to store and purify rainwater, and it is generally used in low-lying areas. Rainwater garden construction has the characteristics of simple maintenance, low capital investment and strong appreciation, so it is more popular with people. (4) Constructed wetlands mainly refer to artificially constructed ground similar to swamps, through which soil, plants and microorganisms are used to treat sewage and sludge. Constructed wetlands play an important role in the construction of Sponge City, which can regulate the water environment of the city and improve the functions of water storage, water purification and flood discharge. Therefore, based on the concept of Sponge City, constructed wetlands should be effectively constructed to purify water sources, protect water bodies and improve water quality.

With the rapid development of cities, more problems have emerged, which requires more efficient and intelligent urban management process. In the allocation process of urban road greening plants, we should not only consider the current situation of the city but also combine the trend of the times to reasonably select and optimize road greening plants. The water resistance of road greening plants is only a short-term water resistance. Hydrophytes that can only grow in an aquatic or hygrophytic environment are not suitable for road greening, because most of the objective conditions of road greening are difficult to meet such a growth environment.

19.2.2 Configuration Mode of Urban Greening Plants

The first consideration of urban road greening plant configuration is the integrity of greening plants. Trees, shrubs, ground cover plants, plants, flowers and plants are reasonably matched to form a plant community with scattered layers and rich colors. In specific operation, it can be considered that street trees are generally single trees or single rows in urban road greening, and the green belt is widened twice. For wetland parks, the most important thing in plant landscape configuration is to follow the principle of ecological protection. The utilization and transformation of urban wetlands cannot be at the cost of sacrificing the balance of the ecosystem. The original ecosystem must be protected to the maximum extent to ensure the stable exchange of material and energy between plants. The purification of water quality depends on the reasonable collocation of aquatic plants, and the surrounding protective layer is served by terrestrial plants. According to local conditions, wetland planners planted a protective forest belt around the wetland to form a barrier, which can also serve as the background of the landscape in the park; Submerged, floating and emergent plants in the park form an interesting contrast. Shrubs, small trees and herbs with strong roots are on the slope protection and river bank, which can prevent water and soil loss and provide public viewing. Sunken green space can reduce surface runoff and urban waterlogging. Rainwater collected by it can penetrate into soil through vegetation layer and permeable pavement, which can not only increase soil water content, supplement groundwater resources, but also save green space irrigation water.

Urban greening design is one of them. In the design process, we should not only pay attention to the aesthetic characteristics but also pay attention to the ability of rain and flood management. First, on the basis of existing road greening plants, add corresponding plants to roads lacking water storage function and improve water storage and water blocking capacity; Secondly, for areas with high rainfall intensity, it is necessary to increase surface plants and build corresponding drainage measures, and drain the water absorbed and retained by plants to corresponding facilities for other purposes, so as to improve the utilization efficiency of water resources and alleviate the urban water shortage. For example, in the process of plant allocation of urban road greening landscape, local mainstream plants are used as the main landscape to highlight local characteristics, and exotic plants that are more suitable for local growth environment are selected as auxiliary landscapes. At the same time, in the process of color selection, on the basis of highlighting the tone of the main scenery, colorful plants should be selected for embellishment, but it should be appropriate, not too much, otherwise it will have a sense of usurping the host's role and be counterproductive.

19.3 Index System of Plant Landscape Configuration in Sponge City Based on Deep Learning

19.3.1 Quantification of Plant Color Richness Index

In deep learning, the improvement of the performance of convolutional neural network mostly depends on increasing the depth and width of the network. Usually, the most direct way to improve the network is to enhance the model by increasing the number of hidden layers and neurons in each layer. However, with the increase of network depth, there will be many problems: larger parameter space, easy over-fitting and more computing resources; The deeper the network, the easier it is for the gradient to disappear and the more difficult it is to optimize. The network structure of Inception v3 is to keep the traditional convolution unchanged at the bottom layer

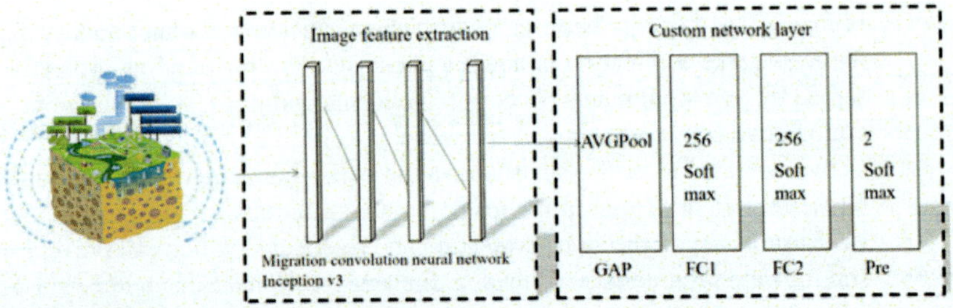

Fig. 19.2 Classification perception model

and connect several improved Inception modules in series at the top layer. The main design criterion of Inception v3 module is to replace the 5*5 convolution kernel with two 3*3 convolution kernels by decomposing convolution, which can dissociate the parameters, so the training speed will be faster. By comprehensively considering the size, performance and accessibility of the model, the Inception v3 model trained based on Image Net dataset under the keras framework is selected for migration. Because the data set in this paper is different from the Image Net data set and there are few data sets, the feature extraction layer is directly migrated. Using pre training weights, a global average pooling layer GAP is customized based on the removal of the top full connection layer in Inception v3 to simplify the operation of a large number of parameters. A full connection layer FC1 with 256 neuron nodes and a rele activation function is used A full connection layer FC2 with 128 neuron nodes and relu activation function, and a full connection layer Pre with 2 output dimensions and softmax activation function are combined to get the final framework of the two category awareness model as shown in Fig. 19.2. In the experiment, only the last four customized network layers are fine tuned.

The scene parsing process based on semantic segmentation involves target recognition and classification in images. Its goal is to assign a category label to each pixel. Because it can predict the label, position and shape of each element, it can provide a complete solution to the scene through scene parsing. Use depth learning to perform semantic segmentation on street view pictures, that is, input image data, and the decoder integrates the environment, surrounding elements and the characteristics of the graph itself to achieve intelligent segmentation of image elements. Because natural images also contain other non-plant elements' color information, such as green trash cans, in order to reduce the interference of these non-plant objects' colors, it is necessary to process other non-plant elements in the original natural images in order to calculate the color value of plants by extracting only the plant elements from the natural images. Before introducing the quantitative method of plant landscape level diversity, this paper first introduces the measurement methods of diversity in most researches. According to the number of categories, diversity can be divided into within the same category and between different categories. The most commonly used methods to measure diversity within the same category are standard deviation and variance, which reflect the degree to which a group of data belonging to the same type deviates from its mean value. The diversity between different categories mainly depends on the number of types and the uniformity of distribution. In most cases, calculating the number of types is the simplest and direct method to measure diversity, such as species richness in ecology and the distribution of species population size. The generalized entropy is the most widely used diversity index. It measures the number of categories while taking into account the uniformity of information distribution. The formula is defined as follows:

$$G_N^a(p_1, p_2, ..., p_N) = (2^{1-a} - 1)^{-1}\left[\sum\nolimits_{i-1}^{N}(p_i^a) - 1\right] \quad (19.1)$$

In which a ≥ 0. When a = 0, it is the richness index; When a approaches 1, it is Shannon entropy index; When a = 2, it is Simpson index, which is used to reflect the balance between categories. The corresponding function form is as follows:

(1) Richness index (a = 0):

$$G_N^a(p_1, p_2, ..., p_N) = N \quad (19.2)$$

(2) Richness index ($a \to 1$):

$$G_N^a(p_1, p_2, ..., p_N) = -\sum\nolimits_{i=1}^{N} p_i \log_2(p_i) \quad (19.3)$$

(3) Abundance index (a = 2):

$$G_N^a(p_1, p_2, ..., p_N) = 1 - \sum\nolimits_{i=1}^{N} {p_i}^2 \quad (19.4)$$

In this paper, the diversity of plant landscape hierarchy refers to the hierarchical structure formed by different community types, which reflects the richness and balance of four hierarchical structures of trees, shrubs, ground cover and flowers in the vertical space of plant communities. It is one of the visual perception factors that affect the quality of urban greening. Based on the generalized entropy function, the richness index, Shannon entropy index and Simpson index were calculated to measure the diversity of plant landscape levels;

$$\text{Level Diversity (simpson)} = 1 - \sum\nolimits_{i=1}^{N}\left[\frac{P_i}{P}\right]^2 \quad (19.5)$$

Among them, N refers to the number of four different plant community types in a segmented Baidu Street View image, including trees, vegetation, grass and flowers, with possible values of {1, 2, 3, 4}. In order to facilitate better understanding and subsequent calculation, for the types and layers of plant landscape, i, Define $i = 1$ for trees, $i = 2$ for vegetation, $i = 3$ for grass, and $i = 4$ for flowers. P_i is the proportion of the number of pixels of the plant community represented by the i category in the segmented image to the total number of pixels in the whole streetscape image, and P is the sum of P_i, that is, $P = P_1 + P_2 + P_3 + P_4$.

Explanation of richness index: The larger the N is, the more levels of plant communities in the vertical direction are and the richer the community types are.

19.3.2 Model Effect Evaluation

As an indispensable part of Sponge City, it should also follow certain principles in its selection. Through investigation, comparison, summary and analysis of plant species selection, allocation and existing problems in Sponge City of Fengxi New Town, Xixian New District, this paper puts forward some relevant principles of plant species selection and allocation in Sponge City. The choice of plant species in Sponge City is suitable for the planting environment, which can ensure the normal growth and function of plants. Therefore, when selecting plants, native plants should be the main ones to ensure their survival and function. At the same time, the cost of later maintenance or replacement of tree species can be reduced, and the construction and management costs can be saved. Giving priority to perennial species and reducing the frequency of plant replacement can not only cooperate with the long-term implementation of LID facilities, but also save labor and cost. Priority should be given to plants that adapt to the site environment, such as those planted in road green belts, and priority should be given to saline-alkali tolerant plants, such as broom grass and Kochia scoparia, to resist salinization of soil. When planting plants in Sponge City, it is necessary to arrange the plants according to different LID measures, which should be accurate and not perfunctory. For example, the problems mentioned in the previous chapter, there are no plants in the transmission grass ditch, so how to give full play to the effects of plants? Therefore, in the plant construction in Sponge City, it is necessary to closely combine plants with LID engineering measures to maximize the plant benefits.

This experiment is designed based on the fine tuning method of transfer learning. Using the pre-trained model on the Image Net dataset, freeze the pre-training model Inception v3 to remove the weights of other convolution layers except the top full connection layer, and customize a pooling layer and three full connection layers based on the removal of the top full connection layer. Based on the partially labeled dataset, only the last four customized convolution layers are subject to parameter fine tuning. In order to obtain the binary classification awareness model, and verify the model effect on the validation data set. The settings of model training super parameters are shown in Table 19.1.

The data in Table 19.1 correspond to the left image name, the right image name and the comparative labeling result value in the labeling tool in Fig. 19.3 in turn. "Left" corresponds to ">", which means that the diversity and health of plants in the left image is obviously better than that in the right image, and "right" corresponds to "<". The original size of the static streetscape image downloaded in the experiment is 4096 × 1536. Therefore, it is necessary to process the size of the image before

Table 19.1 Hyper parameters of binary perception model training

Training super parameters	Value
Batch size	64
Learning rate	1e−4
Randomly inactivated dropout	0.25
Iterations epochs	30
Num classes	2

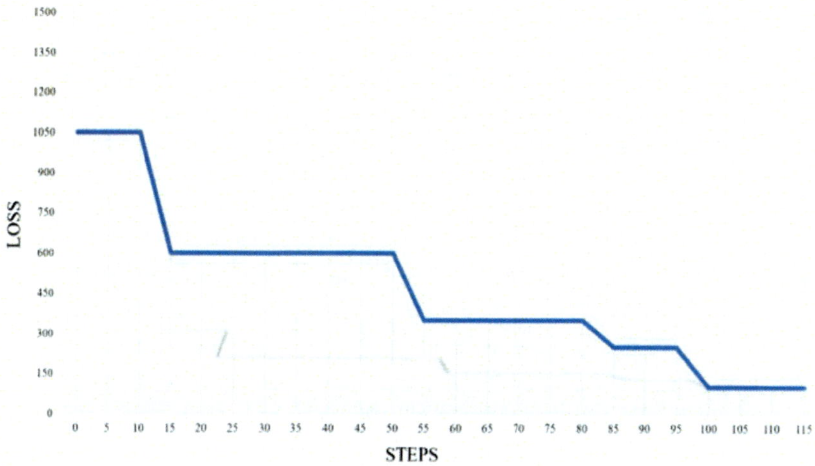

Fig. 19.3 Accuracy curve of training set of classification perception model

inputting the model, and process the static streetscape image into the size required for data input in the model. Considering that the perceptual evaluation of this index is also based on the overall elements contained in the image, the experimental part only adjusts the size of the image by scaling, but not cropping.

The validation set data of labeled samples are input into the model to verify the effect of the model. Finally, the precision and loss function curves of the training set and validation set of the model are shown in Figs. 19.3 and 19.4, respectively. All kinds of curve images are drawn by tensorboard.

Based on the calculated data, the Pearson correlation coefficient of comprehensive evaluation index values among sponge cities is analyzed to prove the visual similarity of the overall plant landscape among urban areas. The evaluation model of plant diversity and health degree is trained based on Baidu static streetscape annotation data set, and the training set loss function, verification set loss function and precision change curve of perception model are obtained after 500, 1000 and 1500 times of training. In addition, in order to verify the actual performance effect of the model, all the static streetscape images in the study area are sorted and scored. The picture shows a group of examples of the change of perception of plant health status from bad to good in streetscape images and the corresponding scoring results. From the

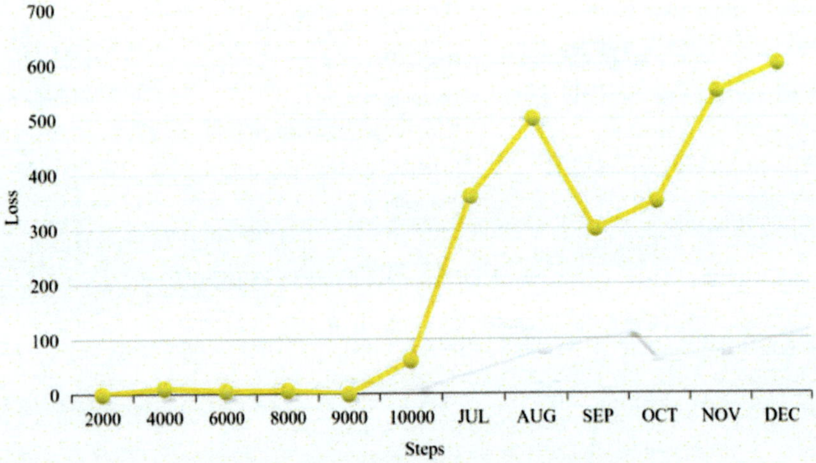

Fig. 19.4 Loss curve of training set of binary perception model

evaluation results, it can be seen that the scoring results of plant diversity and health are generally consistent with the plant growth potential, leaf color and integration with the surrounding environment, etc., and the streetscape images with low and high scoring results are obviously compared with each other, which is obviously representative.

In this experiment, the subjective model of plant diversity and health perception was constructed with reference to RSS-CNN structure, and finally the scoring model structure was constructed. The model is mainly composed of two parts: image feature extraction (transfer learning) and ranking and scoring (ranking learning). In the second part, a ranking network is attached to achieve the perception and scoring of streetscape images. The scoring network is composed of a fully connected network, which is realized based on the comparison results of each group of image pairs. The difference between this experiment and RSS-CNN model is that the first part does not classify each group of images, but directly uses the results of sorting and scoring as a whole to judge the scores of all images, which already includes the process of judging the comparison results of each group of images, which simplifies the network structure compared with RSS-CNN model. The relationship between the comprehensive evaluation index value and the road function level shows that the main body of urban road greening construction in the study area is along the urban general roads and secondary roads. Because there are relatively few buildings and facilities around the road, the construction level of greening on both sides of the road is not bad in theory, but the comprehensive evaluation index results included in these road sections are at a general level, which means that the quality of road plant greening is not only related to the function of the road, but also affected by other factors. Secondly, based on the calculated comprehensive evaluation index of urban road plant landscape, the overall greening quality of each urban area and the visual similarity between urban areas are also analyzed in urban areas. Finally, the internal relationship between the comprehensive evaluation index value and the road function grade is analyzed. The results show that there is a certain correlation between the greening quality of urban road plants and the type of road functions, and more attention is paid to the planting, maintenance and other aspects of plant landscapes on the main urban roads.

19.4 Conclusion

Based on the rapid development of the city, the concept of Sponge City has attracted increasing attention from all walks of life in the process of urban construction, and it is an important concept of urban construction today. The construction of Sponge City requires designers to master all kinds of design knowledge and practical experience, and plant landscape configuration is one of the important aspects, showing its indispensable position more and more. Therefore, for green space facilities in different "sponge cities", we should adjust measures to local conditions and select appropriate "sponge" plants. We should not only create a pleasant urban landscape, but also combine plant planting with urban drought, waterlogging and other environments. When the rainy season comes, we should release and use the stored water resources under the premise that the city urgently needs water resources through water absorption, water storage, water seepage, water purification and other related work, The negative impact caused by the bad urban environment will be greatly alleviated, and the concept of urban greening construction will be brought into full play to ensure a good ecological environment. The theory and practice of this research show that WEDED learning mode can effectively promote deep learning. However, there are still some shortcomings, such as the lack of more practical tests and application strategies. It is believed that through further practice and research and improvement, this model can play a greater role in promoting deep learning.

References

1. Li, Q., Wang, F., Yu, Y., Huang, Z., Li, M., Guan, Y.: Comprehensive performance evaluation of LID practices for the Sponge City construction: a case study in Guangxi, China. J. Environ. Manage. **231**, 10–20 (2019) (In Chinese)
2. Jiang, C., Li, J., Hu, Y., Yao, Y., Li, H.: Construction of water-soil-plant system for rainfall vertical connection in the concept of Sponge City: a review. J. Hydrol. 127327 (2021)
3. Wang, H., Mei, C., Liu, J.: Systematic construction pattern of the Sponge City. J. Hydraul. Eng. **48**(9), 1009–1014 (2017)
4. Zhao, B., Du, Y., Ren, L., Wang, J.: Preparation and performance of epoxy resin permeable bricks for Sponge City construction. J. Appl. Polym. Sci. **137**(34), 49008 (2020) (In Chinese)
5. Yang, P., Zhao, Z.X., Li, Z.C., Wang, Y.H.: Experimental study on long-term performance of new urban green space soil for Sponge City construction. Urban For. Urban Greening **58**, 126906 (2021)

6. Wang, Y., Liu, Z., Wang, G., Xue, W.: Cellular automata based framework for evaluating mitigation strategies of Sponge City. Sci. Total Environ. **796**, 148991 (2021)
7. Chen, L., Wang, Y.: Study on plant configuration and planning of landscape architecture in coastal cities. J. Coast. Res. **115**, 17–20 (2020) (In Chinese)
8. Guo, F., Gao, Z.: Sponge City plant planning and urban construction based on high-resolution remote sensing images. Arab. J. Geosci. **14**(12), 1–15 (2021)
9. Helliwell, M., Thomas, E.J., Whitehouse, D.L.: Reduktion der Lärmbelästigung durch Geschwindigkeitsmessanlagen? Eine Interventionsstudie in Lübeck. Das Gesundheitswesen (Deutsch) **22**, 13563 (2017)
10. Yin, D., Chen, Y., Jia, H., Wang, Q., Chen, Z., Xu, C., et al.: Sponge City practice in China: a review of construction, assessment, operational and maintenance. J. Clean. Prod. **280**, 124963 (2020) (In Chinese)

Chapter 20
Research and Application of Interactive Power Distribution Topology Technology for Distributed New Energy

Gang Wang, Aihua Zhou, Xiaofeng Shen, Sen Pan, Yiqing Wang, Lin Peng, Min Xu, and He Wang

Abstract This paper analyzes the requirements of new energy interactive power distribution topology editing, studies the graph database technology for distributed new energy, designs the power topology function scheme, proposes the interactive power distribution topology mapping technology scheme for new energy, and designs the power topology function framework. The technical scheme proposed in this paper can realize the automatic topology construction, intelligent layout and cabling of the distributed new energy power system, and improve the complex network drawing ability and service level of the topology editing tool.

20.1 Introduction

The interactive power distribution topology analysis for distributed new energy is the key link of power system analysis [1, 2]. With the rapid development and wide application of computer technology, numerical analysis, graph theory, topological analysis, and other computer related mathematical branches have also been fully developed in this field [3, 4]. As a result, the analysis of power system is refreshing, and the form of its mathematical expression, the method of establishing mathematical models, and the corresponding calculation and processing methods have changed greatly. It is precisely this change that meets the requirements of rapid analysis of online real-time control in modern large-scale power systems [5, 6].

G. Wang (✉) · A. Zhou · S. Pan · L. Peng · M. Xu · H. Wang
State Grid Smart Grid Research Institute CO., LTD, Nanjing 210003, China
e-mail: 26261045@qq.com

State Grid Key Laboratory of Information & Network Security, Nanjing 210003, China

X. Shen
State Grid Shanghai Electric Power Company, Shanghai 200122, China

Y. Wang
State Grid Jiangsu Electric Power Co., LTD, Nanjing 210024, Jiangsu, China

S. Patnaik et al. (eds.), *3D Imaging—Multidimensional Signal Processing and Deep Learning*, Smart Innovation, Systems and Technologies 348,
https://doi.org/10.1007/978-981-99-1145-5_20

The topology of interactive power distribution for distributed new energy is complex, which requires high flexibility and ease of use of topology editing tools. The existing grid topology editing tools have cumbersome processes and poor user experience when drawing complex network topology [7, 8]. Design and develop users' habits oriented power distribution and distributed new energy topology editing tools, realize equipment drawing and topology hooking through UI interaction, and improve the complex network drawing ability and service level of topology editing tools through automatic relationship building and intelligent layout and cabling [9, 10].

20.2 Quantitative Control Dynamic Analysis Technique

At present, in the direction of storage, management, visualization and analysis of large-scale networks, the processing capacity of graph database technology is superior to other databases. With the increasing proportion of distributed new energy in the distribution network, it is found that the equipment information data of the distribution network has the characteristics of the network and is suitable for storing some information with graphs in the process of studying the network characteristics of the distribution network topology. From the view of diagram structure, terminal sites such as distributed power supply, distribution station and smart meter can be regarded as nodes in the diagram, the connection between devices can be regarded as the edge between nodes, and the direction of the connection can be regarded as the attribute of the edge. Interactive power allocation for distributed new energy will generate massive monitoring data. If real-time status monitoring is required, such as analyzing the outage range, it will need to query the search algorithm for power range analysis. Traditional database technology can no longer meet these real-time requirements, such as connectivity detection between power grid equipment and the inspection of the relationship between equipments. Therefore, the application of graph database technology greatly meets the requirements of grid equipment storage, query, and analysis, thus improving the efficiency of grid topology management and ensuring the safe operation of the grid [11, 12].

Compared with traditional modeling methods, graph database modeling has higher scalability, flexibility, and availability. Using graph database can easily establish a high fine-grained data model, which is suitable for sorting out various complex relationships. For the construction of knowledge base, it is very convenient to mine data rules from it, which greatly improves the efficiency of data processing. Common methods of graph data modeling include Neo4j graph data modeling, GraphX graph data modeling, TigerGraph graph data modeling, etc. The power grid has typical topology characteristics. The establishment of equipment mapping table based on the topology of the power grid is the key to the integration of power grid data. Neo4j constructs the connection mode between standby nodes by defining three construction units: node, edge, and attribute. In the TigerGraph database, you need to define a Graph Schema before importing data into the data warehouse. A GraphSchema can be considered as a data dictionary that defines different entities (edges and points) and their relationships in a graph.

This paper uses Neo4j open source graph database. Neo4j is a Java-based graph database. It is not only compatible with ACID features, but also has good performance, strong scalability, and high data reliability and security. The strong scalability of Neo4j enables it to process very large-scale complex relational graph data on one machine, and supports parallel processing of multiple machines. Compared with other relational data processing technologies, Neo4j has better performance to process a large number of complex, related, and sparse structured data. Because of its unique graph data model, Neo4j can traverse the graph data without depending on the data size. It traverses the data very quickly, and it also has powerful data processing technology, which is superior to traditional Relational Database Management System (RDBMS) technology. Neo4j carries out information modeling by defining nodes, edges, and attributes. A node may have connections in different directions, and each connection has customizable attributes. Compared with traditional relational databases, graph databases are also relatively simple to create and maintain data. Each node can be understood as a record in a relational database, and edges and their attributes can be understood as fields in this record. Each record can have multiple fields of edges and their attributes. The contents of records can change dynamically. Adding and deleting do not affect the original data logic. Especially for the power grid data in the power system, in general, after the power system has established the graph data model, the association structure of most modules is fixed. However, due to the particularity of the distribution network data, once the association structure changes, the data on the structure will also change, which will often increase the difficulty of database processing, and the database performance will decline due to the change of this structure, In severe cases, the storage results may also be fragmented, and the performance may be further degraded. For this transformation, Neo4j only needs to modify the connection mode between nodes and the attributes of connection dependencies, which can effectively alleviate the performance degradation caused by the change of this result.

20.3 Power Distribution Topology Function

The design concept of separating graphics from equipment is adopted to separate the graphical representation of the equipment used in the topology from its equipment attributes. Edit graphic elements and equipment attributes separately in the graphics window, and closely associate them through the settings of relevant attributes of graphics and equipment. The drawing of station graphics in the graphics window is to show the distributed new energy equipment included in the topology in the form of graphics. Only the color, shape, size, and other graphics are edited here. The tool provides the function of editing device attributes in the device window and the function of editing attributes in the property page. Specific functions include:

(1) The editing function of graphical interface and equipment interface can add, delete, modify, and copy specified elements or equipment.
(2) Read and write EXtensible Markup Language (XML) files and binary files, and provide different safe data storage methods for different requirements of distribution networks.
(3) Edit the device topology relationship by simply dragging the mouse.
(4) The function of associating graphics with equipment. After separation, graphics and equipment are associated through related attribute settings.
(5) Integrity consistency check.
(6) Automatic layout function.
(7) Topology temporal management function.

Integrity consistency check includes checking the uniqueness, correctness, and non-conflict between the graphic unit and the device unit, mainly based on the consistency between the device list information generated by the graphic and the system static device data, the integrity of the topology, and the association between the topology and the device. The software automatically detects the completeness of the data through the integrity consistency check function, and prompts whether the equipment line data is complete by displaying the check results, thus solving the above problems well.

The automatic layout is to automatically generate the topology diagram according to the topology relationship of distributed new energy and distribution network equipment. According to the content of graph theory in data structure, there are generally two ways of graph search, namely depth first search and breadth first search. In the search algorithm described in this paper, the depth first search strategy is adopted. If breadth first search is adopted, it is not known whether the extension direction is the direction of the target device when each distribution network branch is expanded, which results in many expanded branches and large storage capacity. However, depth first search can avoid this problem. It will search along a single path, and will not choose another path to search until the target device cannot be found on this path. In addition, it is not necessary to save the path of the target device that has been searched but cannot be found, which can save storage space.

20.4 Topological Drawing of Power Distribution System

For the modeling of distributed new energy interactive power distribution system, different modeling methods need to be considered according to different business types. The modeling attributes of power distribution system equipment include: power distribution equipment information, network information, and business information. Equipment information includes equipment name, equipment category and subcategory, equipment model, etc. The network information includes IP address, routing list and related configurations. Business information includes the scale of data collection and the business type of data. The graph data modeling can more accurately

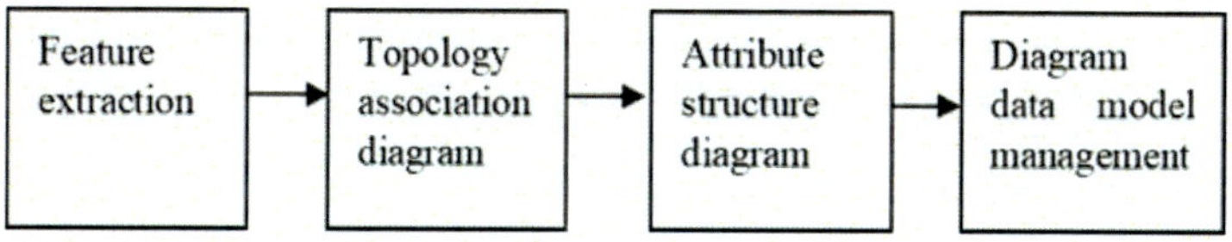

Fig. 20.1 Topology mapping flow chart

distribute the interactive power distribution system of new energy. Different devices in the system often correspond to different power distribution businesses. Different businesses have different requirements for data security and scale. Different priorities can be set for transmission data according to different requirements. For example, data related to power grid security control is superior to other types of business data for transmission.

The equipment in the grid topology data can be used as nodes, and the specific physical connection relationship between equipment and equipment can be seen as edges. The graph data model construction includes four steps: data acquisition, data processing, data import, and data management. The flow chart is shown in Fig. 20.1.

(1) Feature extraction. Feature extraction is mainly to extract the attribute features of the collected distributed power topology data. According to the analysis experience of power system, set the priority of attribute features, arrange the unordered attribute values in descending order according to the priority of influence degree, and then map the obtained attribute information of both continuous and discrete power data into numerical type. Each attribute feature is recorded to generate the first attribute structure topology diagram $G1$. The first attribute structure topology diagram shall at least include device ID, subcategory, device model, IP address, port information, and storage location information. In this way, we can get a set of attribute features that are easy to handle and have power system priority.
(2) Topology association diagram. For the obtained first attribute structure topology diagram $G1$, Pearson product moment correlation coefficient method is used to judge the correlation between equipment and form a topology correlation diagram. Based on Pearson product moment correlation coefficient method, R coefficient is used for correlation analysis, and a strong connection attribute is added to the node with strong correlation to provide a fast channel for distributed power control. Pearson product moment correlation coefficient method is widely used to judge whether there is correlation in data. It is a classical similarity calculation method. It has the characteristics of simple calculation method, low complexity, fast calculation time, and accurate calculation results.
(3) Attribute structure diagram. It is mainly a step to input the generated topology association diagram into the diagram database system, and set the attributes of terminals and edges at the same time. If the connection of the physical wires of the distributed power supply is reflected in the power topology diagram, it is called physical connection. If there is not only physical connection between electrical physical equipment, but also electricity can be connected,

it is said that there is an electrical connection. Generally, the power equipment controls whether the equipment with physical connection is electrically connected through the switch type equipment. In addition, the physical connection is connected but the electrical connection is not connected due to equipment failure and other reasons. After distinguishing between physical connection and electrical connection, set terminals to store the relationship attributes between devices, and form the second attribute structure topology diagram *G*2.

(4) Diagram data model management. It is mainly to combine the business data information with the business priority, and connect the equipment node information to form a unit structure to form a unit framework with different priority levels. Each unit framework is connected. Through the integration of the attributes of equipment nodes and terminals with the topological relationship of the whole system, the third attribute structure topological diagram *G*3 is formed. The refined logic model and topology of the power grid diagram are combined with the three attribute structures to connect the data of the whole diagram and map them to the data model of the power grid topology management diagram. Load historical data and the latest data to realize the real-time update of power grid topology data in the power grid topology database.

20.5 Functional Framework of Interactive Power Distribution Topology for Distributed New Energy

In order to complete the migration of grid topology data from traditional database to graph database, this paper designs the functional framework of power distribution topology according to the characteristics of graph data of grid topology management and the characteristics of graph database itself, as shown in Fig. 20.2.

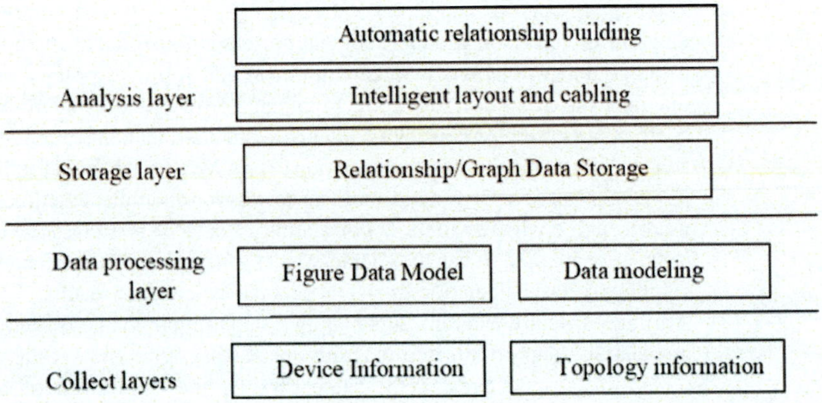

Fig. 20.2 Functional framework of interactive power distribution topology for distributed new energy

(1) Collect layers. The role of this layer is to collect the power topology data of the distribution network, which not only includes the data of physical devices, such as buses, distribution stations, transmission lines, transformers, distributed generation, but also includes the physical connection and topology relationships of accompanying physical devices. Collection of historical grid equipment account information, equipment operation data, equipment model, fault, operation record, and other historical data will help to conduct more comprehensive and effective grid business analysis and improve management efficiency.
(2) Data processing layer. The logical relationship of topology data collected from the power grid is complex, so it is necessary to model graph data based on the data collected from the data collection layer, such as mapping device information to nodes, and the physical connection between devices to edges. At the same time, because there are many redundant and missing data, it is necessary to clean and standardize the data. The main work of this paper is to construct the graph database model.
(3) Storage tier. The standardized data has a graph logic structure, which can be stored through the storage model. Common graph storage data structures include adjacency matrix and adjacency table, and of course, the developed graph database can also be used for storage.
(4) Analysis layer. The research focus of graph database technology for power grid topology management is to analyze graph data. This layer provides some algorithm frameworks to meet the daily power grid needs. It carries out application analysis on power grid topology data to support business distribution and dispatching. Functions include automatic topology construction, intelligent layout and cabling, grid topology search analysis, equipment connectivity analysis, power outage range analysis, etc. The algorithms used include graph traversal algorithm, path search algorithm, etc.

20.6 Conclusions

This paper analyzes the requirements of distributed new energy interactive power distribution topology editing, studies the graph database technology for distributed new energy, designs the power distribution topology function scheme, proposes the interactive power distribution topology mapping technology scheme for distributed new energy, and designs the power distribution topology function framework. The technical scheme proposed in this paper can realize the automatic topology construction, intelligent layout and cabling of the distributed new energy distribution system, and improve the complex network drawing ability and service level of the topology editing tool.

Acknowledgements This work was financially supported by the science and technology project to State Grid Corporation "Research on New Distribution Network Data Deep Fusion and Resource Sharing Technology (5400-202258431A-2-0-ZN)".

References

1. Feng, R.H., Zhao, Z., Xie, S., Huang, J.L., Wang, W.: Low voltage distribution network topology identification method based on principal component analysis and convex optimization. J. Tianjin Univ. **54**(7), 746–753 (2021). (in Chinese)
2. Li, Y.Z., Zhao, Q.S., Liang, D.K., et al.: Construction of low-voltage distribution network impedance topology model based on measurement information. Power Grid Clean Energy **37**(4), 15–22 (2021). (in Chinese)
3. Yan, Y.H., Zhou, R.X., Yu, W., et al.: Station topology identification based on carrier wave and power frequency distortion technology. Autom. Technol. Appl. **40**(10), 68–71 (2021). (in Chinese)
4. Yan, W.G., Wang, G.M., Lin, J.K., et al.: Verification method of low-voltage distribution network topology based on AMI measurement information. China Power **52**(2), 125–133 (2019). (in Chinese)
5. Que, P.X., Gong, X.R., Song, Z.Y, et al.: A complex network topology editing and visualization system based on web: cn110929201a (2020).
6. Zhu, Y.H., Liu, S.X., Ji, L.X., et al.: A sequential link prediction algorithm integrating local topological influence. J. Electron. Inf. **44**(4), 13 (2022)
7. Lei, Y.: Feudal lake topological optimization of material nonlinear substructure based on parametric level set method. Appl. Math. Mech. **42**(11), 11 (2021)
8. Lou, W.J., Wang, L., Jiao, T.C., et al.: Voltage stability preventive control method based on network topology optimization. China Southern Power Grid Technol. **16**(1), 8 (2022)
9. Yao, Y., Xu, J.Y., Cui, Q.Y.: Communication topology optimization of distributed economic dispatching in active distribution network. Energy Eng. **2**, 5 (2020)
10. Shan, S.R., Wang, J.L.: Topology optimization method of wireless monitoring network based on GSO algorithm. Smart Power **48**(6), 7 (2020)
11. Weng, Y.X., Ma, W.Z., Shi, J., et al.: Auxiliary decision-making of power grid operation mode based on semi dynamic topology optimization. Power Eng. Technol. **40**(4), 8 (2021)
12. Li, Z., Chang, V., Ge, J., Pan, L., Hu, H., Huang, B.: Energy-aware task offloading with deadline constraint in mobile edge computing. EURASIP J. Wirel. Commun. Netw. **1**, 1–24 (2021)

Chapter 21
On the Variety of Semilattice-Ordered Semigroup Satisfying X + yxz ≈ X

Hui Xu, Jing Tian, and Junqing Feng

Abstract Basing on a class of automaton formal language, we discuss the variety of semilattice-ordered semigroup satisfying $x + yxz \approx x$, and the result of this paper belongs to the mathematics in multidimensional signal processing. Some identities which can be satisfied by **SLOS**($x + yxz \approx x$) are given. Also, three subvarieties of **SLOS**($x + yxz \approx x$) satisfying the additional identity $x + xy \approx x$, $x + yx \approx x$, and $xz + xyz \approx xz$ are discussed.

21.1 Introduction and Preliminaries

The semigroup theory is an important branch of algebra, and it has important applications in signal processing, information science, automatic control, and so on. The semilattice-ordered semigroup that is a special semigroup with an order relation is closely related to automata theory and formal language. Thus, the study on semilattice-ordered semigroup has important theoretical research value in theoretical computer science. Basing on a class of automaton formal language, we discuss the subvariety of the semilattice-ordered semigroup, and the result of this paper belongs to the mathematics in multidimensional signal processing.

A structure $(S, \cdot, \vee)$ is called a semilattice-ordered semigroup if

(i) $(S, \cdot)$ is a semigroup;
(ii) $(S, \vee)$ is a semilattice;
(iii) $(\forall a, b, c \in S)a(b \vee c) = ab \vee ac$ and $(a \vee b)c = ac \vee bc$.

H. Xu (✉) · J. Feng
Fundamentals Department, Air Force Engineering University, Xi′an, Shaanxi 710043, Republic of China
e-mail: xaxuhui716@126.com

J. Tian
Economy and Finance School, Xian International Studies University, Xian, Shaanxi 710128, Republic of China

S. Patnaik et al. (eds.), *3D Imaging—Multidimensional Signal Processing and Deep Learning*, Smart Innovation, Systems and Technologies 348,
https://doi.org/10.1007/978-981-99-1145-5_21

Denote by **SLOS** the variety of all semilattice-ordered semigroups. The free objects of subvarieties of **SLOS** are widely studied in literatures. Zhao [1] gave models of the free object in the subvariety **SLOS** satisfying the identity $x^2 \approx x$. Further, Pastijn et al. [2–4] determined the lattice of subvarieties of **SLOS** and solved the word problem for the free objects in those subvarieties. Kuřil and Polák [5] provided a construction of the free objects in subvarieties satisfying the identity $x^r \approx x$, where $r \geq 2$ is an integer. As a generalization of results from [5], Ren and Zhao [6] constructed a free object in subvariety of **SLOS** satisfying the additional identity $x^n \approx x^m$ for any positive integers n and m. For related research, please refer to [7–9].

Furthermore, Wang et al. [10] showed that the collection of all subwordfree languages is freely generated in **SLOS** satisfying the additional identities $x + xy \approx x$ and $x + yx \approx x$. Recently, Tian et al. [11] introduced three collections of all independent languages with respect to three binary relations and established the combinatorial properties and algebraic structures of these languages. By equipping them with two binary operations, it is shown that they are freely generated in the subvariety of **SLOS** satisfying the identity $x + xy \approx x(x + yx \approx x, xz + xyz \approx xz$, resp.).

In this paper, we shall discuss the subvariety of **SLOS** satisfying the identity $x + yxz \approx x$, and we get some identities that can be satisfied by **SLOS**$(x + yxz \approx x)$.

Let O_n, L_n, R_n, and I_n denote the identities by

$$O_n : x_1x_2 \ldots x_n + x_1y_1x_1y_2 \ldots x_{n-1}y_{n-1}x_n \approx x_1x_2 \ldots x_n$$

$$L_n : x_1x_2 \ldots x_n + x_1y_1x_1y_2 \ldots x_{n-1}y_{n-1}x_ny_n \approx x_1x_2 \ldots x_n$$

$$R_n : x_1x_2 \ldots x_n + y_1x_1y_2x_2 \ldots y_nx_n \approx x_1x_2 \ldots x_n$$

$$I_n : x_1x_2 \ldots x_n + y_1x_1y_2x_2 \ldots y_nx_ny_{n+1} \approx x_1x_2 \ldots x_n$$

Tian et al. [11] have proved that the variety **SLOS**$(xz + xyz \approx xz)$ (**SLOS**$(x + xy \approx x)$, **SLOS**$(x + yx \approx x)$ resp.) satisfy the identity O_n (L_n, R_n, resp.) Also, we know both **SLOS**$(x + xy \approx x)$ and **SLOS**$(x + yx \approx x)$ satisfy the identity O_n.

In Sect. 21.2, we shall prove **SLOS**$(x + yxz \approx x)$ satisfies I_{2k-1}, L_{2k}, R_{2k}, and O_{2k+1}. In Sect. 21.3, we shall discuss three subvarieties of **SLOS**$(x + yxz \approx x)$ satisfying the additional identity $x + xy \approx x, x + yx \approx x$, and $xz + xyz \approx xz$. We will prove the identities I_n, L_n, R_n, and O_n which are true in **SLOS**$(x + yxz \approx x, xz + xyz \approx xz)$ except for L_1 and R_1.

21.2 Identities Satisfied by SLOS($X + yxz \approx X$)

In this section, we consider the identities satisfied by $x + yxz \approx x$. We shall prove $I_{2k-1}, L_{2k}, R_{2k}, O_{2k+1}$ can be implied by $x + yxz \approx x$.

Proposition 21.1 The variety **SLOS**($x + yxz \approx x$) satisfies the identity $x_1x_2 \ldots x_n + y_1x_1y_2x_2 \ldots y_nx_ny_{n+1} \approx x_1x_2 \ldots x_n$ when n is odd.

If n is even, only when $|y_i| \geq 2$ for some $i \in \{1, 2, \ldots, n\}$ (I_n) is satisfied.

Proof Denote I_1 by

$$x_1 + y_1x_1y_2 \approx x_1.$$

Post-multiply x_2 by I_1 and pre-multiply x_3 by I_1 from both sides,

$$x_1x_2 + y_1x_1y_2x_2 \approx x_1x_2 \tag{21.1}$$

$$x_3x_1 + x_3y_1x_1y_2 \approx x_3x_1 \tag{21.2}$$

Also, we have

$$x_3 + y_3x_3y_4 \approx x_3 \tag{21.3}$$

After multiplying (21.1) and (21.3), we get

$$x_1x_2x_3 + x_1x_2y_3x_3y_4 + y_1x_1y_2x_2x_3 + y_1x_1y_2x_2y_3x_3y_4 \approx x_1x_2x_3.$$

Applying (21.1) and (21.2), we obtain

$$x_1x_2x_3 + x_1x_2y_3x_3y_4 \approx x_1x_2x_3.$$

Then, we get $I_3 : x_1x_2x_3 + y_1x_1y_2x_2y_3x_3y_4 \approx x_1x_2x_3$.

In the remaining part of this work, we shall prove I_{2k+1} is true based on induction on k.

Suppose that I_{2k-1} is true, that is, I_{2k-1}: $x_1x_2 \ldots x_{2k-1} + y_1x_1y_2x_2 \ldots y_{2k-1}x_{2k-1}y_{2k} \approx x_1x_2 \ldots x_{2k-1}$.

We post-multiply both sides by x_{2k},

$$x_1x_2 \ldots x_{2k-1}x_{2k} + y_1x_1y_2x_2 \ldots y_{2k-1}x_{2k-1}y_{2k}x_{2k} \approx x_1x_2 \ldots x_{2k-1}x_{2k}. \tag{21.4}$$

Also, we know that

$$x_{2k+1} + y_{2k+1}x_{2k+1}y_{2k+1} \approx x_{2k+1}. \tag{21.5}$$

After multiplying (21.4) and (21.5) is got:

$$\begin{aligned}&x_1x_2\cdots x_{2k-1}x_{2k}x_{2k+1} + x_1x_2\cdots x_{2k-1}x_{2k}y_{2k+1}x_{2k+1}y_{2k+}\\&+ y_1x_1y_2x_2\cdots y_{2k-1}x_{2k-1}y_{2k}x_{2k}x_{2k+}\\&+ y_1x_1y_2x_2\cdots y_{2k-1}x_{2k-1}y_{2k}x_{2k}y_{2k+1}x_{2k+1}y_{2k+1}\\&\approx x_1x_2\cdots x_{2k-1}x_{2k}x_{2k+1}.\end{aligned}$$

Because

$$x_1x_2\cdots x_{2k-1}x_{2k}x_{2k+1} + x_1x_2\cdots x_{2k-1}x_{2k}y_{2k+1}x_{2k+1}y_{2k+1} \approx x_1x_2\cdots x_{2k-1}x_{2k}x_{2k+1}.$$

$$x_1x_2\cdots x_{2k-1}x_{2k}x_{2k+1} + y_1x_1y_2x_2\cdots y_{2k-1}x_{2k-1}y_{2k}x_{2k}x_{2k+1} \approx x_1x_2\cdots x_{2k-1}x_{2k}x_{2k+1}.$$

Thus, we get

$$I_{2k+1} : x_1x_2\cdots x_{2k-1}x_{2k}x_{2k+1} + y_1x_1y_2x_2\cdots y_{2k-1}x_{2k-1}y_{2k}x_{2k}y_{2k+1}x_{2k+1}y_{2k+1} \approx x_1x_2\cdots x_{2k-1}x_{2k}x_{2k+1}.$$

Summing the above results, we have proved I_{2k+1} is satisfied in **SLOS**($x + yxz \approx x$).

In the following, we consider I_{2k}.

$$x_1 + y_1x_1y_2 \approx x_1. \tag{21.6}$$

$$x_2 + y_3x_2y_4 \approx x_2. \tag{21.7}$$

By multiplying (21.6) by (21.7) is got: $x_1x_2 + x_1y_3x_2y_4 + y_1x_1y_2x_2 + y_1x_1y_2y_3x_2y_4 \approx x_1x_2$.

After applying (21.1) and (21.2), we have

$$x_1x_2 + y_1x_1y_2y_3x_2y_4 \approx x_1x_2.$$

Then, we get

$$x_1x_2 + y_1x_1y_2x_2y_3 \approx x_1x_2$$

where $|y_2| \geq 2$. Next, we shall prove

$$x_1x_2 + \sigma_1\sigma_2x_1\sigma_3x_2\sigma_4 \approx x_1x_2.$$

In fact,

$$x_1 + \sigma_2x_1\sigma_3 \approx x_1,$$

$$\sigma_1 x_1 + \sigma_1\sigma_2 x_1\sigma_3 \approx \sigma_1 x_1.$$

$$\sigma_1 x_1 x_2\sigma_4 + \sigma_1\sigma_2 x_1\sigma_3 x_2\sigma_4 \approx \sigma_1 x_1 x_2\sigma_4.$$

$$x_1x_2 + \sigma_1\sigma_2 x_1\sigma_3 x_2\sigma_4 \approx (x_1x_2 + \sigma_1 x_1x_2\sigma_4) + \sigma_1\sigma_2 x_1\sigma_3 x_2\sigma_4$$
$$\approx x_1x_2 + (\sigma_1 x_1x_2\sigma_4 + \sigma_1\sigma_2 x_1\sigma_3 x_2\sigma_4) \approx x_1x_2 + \sigma_1 x_1x_2\sigma_4 \approx x_1x_2.$$

Similarly, we can prove

$$x_1x_2 + \sigma_1 x_1\sigma_2 x_2\sigma_3\sigma_4 \approx x_1x_2.$$

Summing the above results, we have proved

$$x_1x_2 + y_1x_1y_2x_2y_3 \approx x_1x_2,$$

where some $|y_i| \geq 2$.

In the following, we shall prove that I_{2k} is satisfied when $|y_i| \geq 2$ for some $i \in \{1, 2, \ldots, 2k\}$.

Suppose I_{2k-1} is true.

$$I_{2k-1} : x_1x_2 \ldots x_{2k-1} + y_1x_1y_2x_2 \ldots y_{2k-1}x_{2k-1}y_{2k} \approx x_1x_2 \ldots x_{2k-1}x_{2k-1}.$$

Then

$$\sigma_1 x_1x_2 \ldots x_{2k-1}x_{2k}y_{2k+1} + \sigma_1 y_1x_1y_2x_2 \ldots y_{2k-1}x_{2k-1}y_{2k}x_{2k}y_{2k+1}$$
$$\approx \sigma_1 x_1x_2 \ldots x_{2k-1}x_{2k}y_{2k+1}.x_1x_2 \ldots x_{2k-1}x_{2k}$$
$$+ \sigma_1 x_1x_2 \ldots x_{2k-1}x_{2k}y_{2k+1} + \sigma_1 y_1x_1y_2x_2 \ldots y_{2k-1}x_{2k-1}y_{2k}x_{2k}y_{2k+1}$$
$$\approx x_1x_2 \ldots x_{2k-1}x_{2k} + \sigma_1 x_1x_2 \ldots x_{2k-1}x_{2k}y_{2k+1}.$$

It is routine that

$$x_1x_2 \cdots x_{2k-1}x_{2k} + \sigma_1 y_1x_1y_2x_2 \cdots y_{2k-1}x_{2k-1}y_{2k}x_{2k}y_{2k+1} \approx x_1x_2 \cdots x_{2k-1}x_{2k}.$$

Similarly, we can prove

$$x_1x_2 \cdots x_{2k-1}x_{2k} + y_1x_1y_2x_2 \cdots y_{2k-1}x_{2k-1}y_{2k}x_{2k}y_{2k+1}\sigma_1 \approx x_1x_2 \cdots x_{2k-1}x_{2k}.$$

Thus, we prove I_{2k} is satisfied when $|y_1| \geq 2$ and $|y_{2k+1}| \geq 2$.

Also, we know that

$$x_1 + y_1x_1y_2 \approx x_1. \tag{21.8}$$

$$x_2x_3\ldots x_{2k} + y_2'x_2y_3\ldots y_{2k}x_{2k}y_{2k+1} \approx x_2x_3\ldots x_{2k}. \quad (21.9)$$

Then, multiply (21.8) and (21.9)

$$x_1x_2x_3\ldots x_{2k} + x_1y_2'x_2y_3\ldots y_{2k}x_{2k}y_{2k+1} + y_1x_1y_2x_2x_3\ldots x_{2k} + y_1x_1y_2y_2'x_2y_3\ldots y_{2k}x_{2k}y_{2k+1} \approx x_1x_2x_3\ldots x_{2k}.$$

Applying L_{2k} and R_{2k}, we prove that I_{2k} is true when $|y_2| \geq 2$. By multiplying I_2 and I_{2k-2}, we can prove I_{2k} is satisfied when $|y_3| \geq 2$. Similarly, by applying I_i and I_{2k-i}, we can prove I_{2k} is satisfied when $|y_{i+1}| \geq 2$.

Summing the above results, I_{2k} is satisfied when $|y_i| \geq 2$ for some $i \in \{1, 2, \ldots, 2k\}$.

By post(pre)-multiplying x_{n+1} by I_n, we get R_{n+1} and L_{n+1}, resp. In addition, O_{n+1} can be got by R_n and L_n. It is routine that we can prove the following propositions.

Proposition 21.2 The variety **SLOS**$(x + yxz \approx x)$ satisfies the identity

$$L_n : x_1x_2y_2\ldots x_n + x_1y_1x_2y_2\ldots x_ny_n \approx x_1x_2\ldots x_n \text{ when } n \text{ is even.}$$

If n is odd, only when $|y_i| \geq 2$ for some $i \in \{1, 2, \ldots, n\}$ (L_n) is satisfied.

Proposition 21.3 The variety **SLOS**$(x + yxz \approx x)$ satisfies the identity

$$R_n : x_1x_2\ldots x_n + y_1x_1y_2x_2\ldots y_nx_n \approx x_1x_2\ldots x_n\ R \text{ when } n \text{ is even.}$$

If n is odd ($n \geq 3$), only when $|y_i| \geq 2$ for some $i \in \{1, 2, \ldots, n\}$ (R_n) is satisfied.

Proposition 21.4 The variety **SLOS**$(x + yxz \approx x)$ satisfies the identity

$$O_n : x_1x_2\ldots x_n + x_1y_1x_2y_2\ldots y_{n-1}x_n \approx x_1x_2\ldots x_n \text{ when } n \text{ is odd.}$$

If n is even($n \geq 4$), only when $|y_i| \geq 2$ for some $i \in \{1, 2, \ldots, n\}$ (O_n) is satisfied.

21.3 Three Subvarieties of SLOS($X + yxz \approx X$)

We have known I_1 can imply I_{2n+1}, L_{2n}, R_{2n}, and O_{2n+1}. In this section, we shall prove $I_1 \cup L_1$ can imply I_n, L_n, R_n, and O_n for any integer n except R_1.

Tian et al. [11] proved that L_n (R_n, O_n, resp.) can be implied from L_1 (R_1, O_2, resp.). We now restate three Lemmas from [11].

Lemma 21.1 Tian et al. [11] The variety SLOS$(x + xy \approx x)$ satisfies the identity L_n for any integer $n \geq 1$.

Lemma 21.2 Tian et al. [11] The variety **SLOS**($x + yx \approx x$) satisfies the identity R_n for any integer $n \geq 1$.

Lemma 21.3 Tian et al. [11] The variety SLOS($xz + xyz \approx xz$) satisfies the identity O_n for any integer $n \geq 2$.

Basing on the above Lemmas, we can prove the following propositions.

Proposition 21.5 The variety **SLOS**($x + yxz \approx x, x + xy \approx x$) satisfies I_n, L_n, R_n, (n > 1) and O_n.

Proof Let $|\sigma_i| = 1$. First, we prove

$$x_1x_2 \ldots x_{2k} + \sigma_0 x_1 \sigma_1 x_2 \ldots \sigma_{2k-1} x_{2k} \sigma_{2k} \approx x_1x_2 \ldots x_{2k}.$$

By Proposition 21.1, we know I_{2k-1} and R_{2k} can be implied by I_1. Also,

$$x_{2k} + x_{2k}\sigma_{2k+1} \approx x_{2k}. \tag{21.10}$$

$$x_1x_2x_3 \ldots x_{2k-1} + \sigma_0 x_1 \sigma_1 x_2 x_3 \ldots \sigma_{2k-2} x_{2k-1} \sigma_{2k} \approx x_1x_2x_3 \ldots x_{2k-1}. \tag{21.11}$$

Then multiply (21.10) and (21.11) and apply R_{2k}, we get $x_1x_2x_3 \ldots x_{2k} + \sigma_0 x_1 \sigma_1 x_2 x_3 \ldots \sigma_{2k-2} x_{2k-1} \sigma_{2k} x_{2k} \sigma_{2k+1} \approx x_1x_2x_3 \ldots x_{2k}$.
We post-multiply both sides by x_{2k+1}, then

$$x_1x_2 \ldots x_{2k} + \sigma_1 x_1 \sigma_2 x_2 \ldots \sigma_{2k+1} x_{2k+1} \approx x_1x_2 \ldots x_{2k+1}.$$

On the other hand, I_1 can imply O_2, by Lemma 21.3, O_n can be implied by O_2. Thus, we have proven

$$x_1x_2 \ldots x_{2k} + x_1 \sigma_1 x_2 \sigma_2 \ldots \sigma_{2k} x_{2k} \approx x_1x_2 \ldots x_{2k}.$$

Proposition 21.6 I_n, L_n ($n > 1$), R_n, and O_n can be implied by $x + yxz \approx x$ and $x + yx \approx x$.

Proposition 21.7 I_n, L_n ($n > 1$), R_n ($n > 1$), and O_n can be implied by $x + yxz \approx x$ and $xz + xyz \approx xz$.

21.4 Conclusions

The semilattice-ordered semigroup which has closed relation with automata theory and formal language is studied in this paper. We discussed the subvariety of **SLOS** satisfying the identity $x + yxz \approx x$, and we proved **SLOS**($x + yxz \approx x$) satisfies I_{2k-1}, L_{2k}, R_{2k}, and O_{2k+1}. The result of this paper has important theoretical research value in formal language theory.

References

1. Zhao, X.Z.: Idempotent semirings with a commutative additive reduct. Semigroup Forum **64**, 289–296 (2002)
2. Ghosh, S., Pastijn, F., Zhao, X.: Varieties generated by ordered bands I. Order **22**, 109–128 (2005)
3. Pastijn, F.: Varieties generated by ordered bands II. Order **22**, 129–143 (2005)
4. Pastijn, F., Zhao, X.: Varieties of idempotent. Algebra Universslis **54**, 301–321 (2005)
5. Kuřil, M., Polák, L.: On varieties of semilattice-ordered semigroups. Semigroup Forum **71**, 27–48 (2005)
6. Ren, M., Zhao, X.: On free burnside AI-semirings. Semigroup Forum **90**(1), 174–183 (2015)
7. Ren, M., Zhao, X., Shao, Y.: On a variety of burnside AI-semirings satisfying $x^{\mathrm{n}} \approx x$. Semigroup Forum **93**(3), 501–515 (2016)
8. Ren, M., Zhao, X., Shao, Y.: The lattice of AI-semiring varieties satisfying $x^{\mathrm{n}} \approx x$ and $xy \approx yx$. Semigroup Forum **100**(8) (2020)
9. Cheng, Y., Shao, Y.: Semiring varieties related to multiplicative Green's relations on a semiring. Semigroup Forum **101**(3) (2020)
10. Wang, Z., Liang, F., He, Y., Yang, D.: Semiring structures of some classes of hypercodes. J. Autom. Lang. Comb. **14**, 259–272 (2009)
11. Tian, J., Shao, Y., Zhao, X.: Out subword-free languages and its subclasses. Int. J. Found. Comput. Sci. **27**(3), 305–326 (2016)

Chapter 22
An Application for Color Feature Recognition from Plant Images

Xiang Liu and Zaiqing Chen

Abstract With the development of image processing technology, more and more applications need to identify the color information in the image. For example, in autonomous driving technology, it is necessary to distinguish the color of traffic lights, and in agricultural planting technology, it is necessary to classify the color of crops, etc., so the color feature recognition of images is of great significance, and this technology will be applicable to various scenarios. This paper mainly identifies the color features of plant images. We obtain the standard color comparison card sample set and divide the standard color comparison card sample set into a training set and a test set. A sample set of standard color charts contains (1) the RGB value of the color number and (2) CIE L*a*b* value of color number. Then, the recognition accuracy test is carried out by using the recognition model of the test set, and based on the result of the recognition accuracy test, a target plant color recognition database is established. We obtain the CIE L*a*b* value of the target image, correspond to the database, and get the corresponding color result. The recognition accuracy is 97%.

22.1 Introduction

Images are the basis of human perception of the world. It is also an important way for human beings to obtain, express, and transmit information. Image feature recognition has very important applications in many scenarios. Color diagnosis is probably the most investigated property in studies exploring the role of color information in object recognition [1]. In a target image, there are many important features, including the outline, texture, and color of the object [2]. When performing image processing, color is a feature that cannot be ignored. We need to accurately identify the color to meet

X. Liu · Z. Chen (✉)
Yunnan Key Laboratory of Optoelectronic Information Technology, Kunming, Yunnan, China
e-mail: Zaiqingchen@ynnu.edu.cn

Z. Chen
School of Information Science and Technology, Yunnan Normal University, Kunming, Yunnan, China

S. Patnaik et al. (eds.), *3D Imaging—Multidimensional Signal Processing and Deep Learning*, Smart Innovation, Systems and Technologies 348,
https://doi.org/10.1007/978-981-99-1145-5_22

people's needs. We can even consider a different way of handling color features. This study is about an application of extracting plant color features. This research aims to quickly detect the color characteristics of plants. The traditional manual visual detection method is easily affected by the environment; under different light intensities, the human eye may make a wrong judgment. Compared with the traditional manual visual inspection method, this new detection method is less susceptible to the influence of the environment.

Humans have never stopped exploring colors. People's understanding of color is not entirely determined by the physical properties of light. For example, people's perception of color is often influenced by the surrounding colors. People now often regard color as the basic feature of an object, such as the sky is blue, the leaves are green, and the apple is red, etc. However, modern science has long revealed that the essence of color is the wavelength or frequency of light, and light of different wavelengths will have different colors. The color of the object we see is only the color produced by the object reflecting the natural light outside. But this understanding alone is not enough, because the wavelength of light changes continuously, ranging from gamma rays at the microscopic scale to ultra-long electromagnetic waves at the macroscopic scale, and the wavelength range that humans can recognize is only between 380 and 780 nm. For humans, the light we can see is only part of a very small number of electromagnetic waves. In this very small range of visible light, there are only three colors that we can really distinguish, namely red, green, and blue. These three colors of light can be received by our three different cone cells, which produce different colored signals on our retina. And all the other colors we see are just combinations of these three basic colors. The Munsell color system is a color system widely used today. In 1919, it was subjected to a spectrophotometric examination by John and Arthur [3].

In this study, we used the Munsell color system. It is a very important color system, and many scholars are using this color system to represent colors. The characteristic of this color system is that the color is divided into ten hues. Through these ten hues, people can quickly identify colors.

We will adopt the Royal Horticultural Society (RHS) standard color card, which has 920 colors [4]. The RHS color chart is the standard reference used by horticulture around the world to record plant color. It can precisely match flowers, fruit, and other plants to accurately record and communicate color all over the world. Each color has a number and letter code and a name. The RHS color chart is the standard reference tool used by horticulturists around the world to record plant colors.

At present, when using the standard color card to identify the color of plants, the method of visual comparison is usually used. The human eye compares the different color numbers of the plant material and the color card one by one and finds the color that is closest to the color of the plant material visually. Define this color as the original color of the plant. However, differences in human vision and cognition make the comparison results highly subjective and poor in accuracy. In addition, the environment will also have a greater impact on the color card visual comparison method. In different environments, human eyes may have different judgments on the same color. Therefore, there is a large error in plant color recognition through the color card visual comparison method. This study is of great significance for extracting the color characteristics of plants.

We obtain the standard color comparison card sample set and divide the standard color comparison card sample set into a training set and a test set. A sample set of standard color charts contains (1) the RGB value of the color number and (2) CIE $L^*a^*b^*$ value of color number. Then, the recognition accuracy test is carried out by using the recognition model of the test set, and based on the result of the recognition accuracy test, a target plant color recognition database is established. We need to evaluate the accuracy of the color model [5]. We obtain the RGB value of the target image, convert the RGB value to the XYZ value, then convert the XYZ value to the CIE $L^*a^*b^*$ value, and finally correspond to the database to obtain the corresponding color result.

22.2 Experimental Principle

Color is the element of form that arouses our greatest sensitivity. It is one of the most expressive elements. Color has three basic properties: hue, value, and chroma [6]. According to the characteristics of the human eye, a color image is generally divided into three channels for representation, namely R, G, and B [7].

The XYZ tristimulus value is obtained by linear transformation of the RGB color space, and the transformed space is the CIE XYZ color space, which is equivalent to using the XYZ base of the matching color to replace the RGB base to represent the color. In theory, we can get the RGB value of the plant image and then convert the RGB value to the XYZ value. The transformation matrix is as follows:

$$\begin{bmatrix} X \\ Y \\ Z \end{bmatrix} = \frac{1}{b_{21}} \begin{bmatrix} b_{11} & b_{12} & b_{13} \\ b_{21} & b_{22} & b_{23} \\ b_{31} & b_{32} & b_{33} \end{bmatrix} \begin{bmatrix} R \\ G \\ B \end{bmatrix} = \frac{1}{0.17697} \begin{bmatrix} 0.49 & 0.31 & 0.20 \\ 0.17697 & 0.81240 & 0.01063 \\ 0.00 & 0.01 & 0.99 \end{bmatrix} \begin{bmatrix} R \\ G \\ B \end{bmatrix} \tag{22.1}$$

Manufacturers and designers usually use the CIE $L^*a^*b^*$ 3D color space to quantify the color's properties and determine the numerical difference between shades. The CIE $L^*a^*b^*$ color model is composed of lightness (L^*) and three elements of color a^* and b^*. L^* represents lightness, a represents the range from magenta to green, and b represents the range from yellow to blue [8]. To convert XYZ values to CIE $L^*a^*b^*$ values, the conversion formula is as follows:

$$\begin{cases} L^* = 116 f(Y/Y_n) - 16 \\ a^* = 500[f(X/X_n) - f(Y/Y_n)] \\ b^* = 200[f(Y/Y_n) - f(Z/Z_n)] \end{cases} \tag{22.2}$$

$$f(t) = \begin{cases} t^{1/3} & \text{if } t > \left(\frac{6}{29}\right)^3 \\ \frac{1}{3}\left(\frac{29}{6}\right)^2 t + \frac{4}{29} & \text{otherwise} \end{cases} \quad (22.3)$$

Through the above process, we can get the CIE $L^*a^*b^*$ value from RGB value of the image. These two values are two very important parameters. They can directly and quantitatively represent the color. In nature, there are thousands of colors, and it is difficult for human beings to distinguish them with the naked eye, so we must express colors through standard methods. When we get the value of RGB and the value of CIE $L^*a^*b^*$, we can determine the color.

We express the identified results in the Munsell color system. The Munsell color system is a method in colorimetry to describe colors through three dimensions of value, hue, and chroma [9]. This color description system was created by American artist Albert H. Munsell (1858–1918) in 1898 and adopted by the USDA as the official color description system for soil research in the 1930s [10]. It is still the standard of color comparison. Most studies on color refer to the Munsell color system.

There are a total of ten hues in the Munsell system, and they are red (R), red yellow (YR), yellow (Y), yellow green (GY), green (G), green blue (BG), blue (B), blue purple (PB), purple (P), and purple red (RP) [11]. The identification form of a specific color is hue + value + chroma. For example, 5B 5/10 is a true blue (5B) with medium lightness (5) and high chroma (10). Through this representation of color, people can clearly understand the specific properties of the color.

22.3 Experimental Environment

This research is about the application of color feature recognition in plant images. This experiment is carried out by an image color recognition system, as shown in Fig. 22.1. This system is developed based on the Visual Studio 2022 platform, and the programming language used is C Sharp. This study uses Microsoft Visual Studio Enterprise 2022 (64 bit), and version number is 17.3.4 [12]. The platform has some advantages. It is mainly reflected in the following aspects: (1) industry-leading database tools, (2) efficient architecture guidance, and (3) key test functions. Through this image color recognition system, we can measure the standard color card sample set and get their RGB value and CIE L*a*b* value.

Through Fig. 22.1, people can see the interface of the color analysis system. The main function of this color analysis system is to analyze the color category of plants and extract the color characteristics of plants. This research uses camera equipment to take images of plants then use the color analysis system to process the plant image.

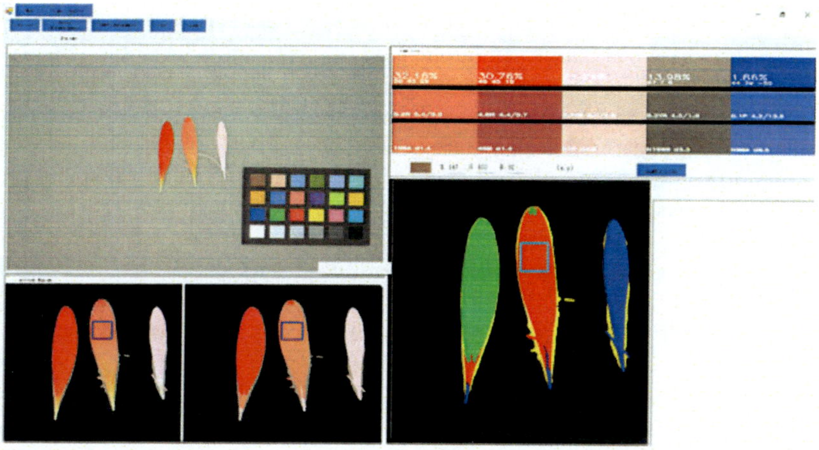

Fig. 22.1 Color analysis system

22.4 Data Collection and Testing of Standard Color Cards

We use the RHS plant color chart to test the recognition accuracy. The RHS color card is an international standard reference for recording plant colors. It can accurately restore the color of plants in nature. Therefore, scholars will use the standard RHS color card in the study of plant color. There are a total of 221 color cards used in this experiment, with a total of 884 color numbers. In this study, 884 color numbers were collected sequentially. A color analysis system will be the tool for dealing with these color cards. Color recognition system is used to collect data in turn to obtain the RGB value and CIE $L^*a^*b^*$ value of each color number. These two values will be important parameters in the study. First, we open the color analysis system then import the plant image and process it. This study will demonstrate the process of processing with a color analysis system. Figure 22.2 is a diagram of the collection process of the color number.

(a) (b)

Fig. 22.2 Color number collection process: **a** color characterization, **b** color feature extraction of image region

We collected each color number in this way and then got 884 data. We will use these 884 data to form a training set. Then use a set of the same color cards for comparison to judge the accuracy of the color recognition system. After a lot of repeated experiments and comparisons, it was found that several color numbers were not accurate enough to be easily recognized as other colors. We identified a total of 884 color numbers. There are 860 correct color numbers and 24 wrong color numbers. We will record the wrong color number and then organize it into Table 22.1.

Through the analysis of these error color numbers, we can get some key data, such as the value of CIE $L^*a^*b^*$. This helps the analysis to identify wrong color areas. When we know these unrecognizable color areas, we can clarify the direction of optimization.

Table 22.1 Identifying the wrong color number

Color number	Wrong recognition result	Value of CIE $L^*a^*b^*$
6D	8C	97/-5/27
18D	13D	98/0/16
21A	17A	83/15/72
21B	15A	85/11/63
21C	16A	89/8/55
21D	20B	92/6/37
23C	22B	90/9/39
28A	N25A	69/40/45
28B	30C	74/37/50
28C	29B	84/22/34
28D	26C	89/15/23
30A	32B	70/44/39
30B	32B	72/43/44
30C	169D	73/38/45
30D	28A	78/32/41
N30A	N34B	58/52/32
N30B	32A	66/51/41
N30C	N25	68/47/46
N30D	30B	72/41/52
44C	42C	62/47/26
45A	46B	47/43/16
50B	47C	63/45/12
N66C	68A	68/45/−11
70D	63D	87/19/−4

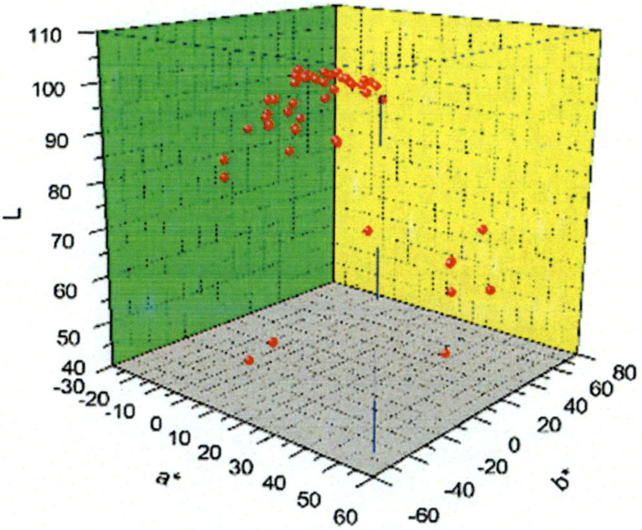

Fig. 22.3 Identifying the wrong color number contains the position in the CIE $L^*a^*b^*$ color space

For these color numbers that cannot be accurately identified, we can draw their positions in the CIE $L^*a^*b^*$ color space as shown in Fig. 22.3.

Through the analysis of Fig. 22.3, we can observe the misidentified color regions. These regions are mainly concentrated in the upper half of the CIE $L^*a^*b^*$ color space. Essentially, the misidentified color number mainly comes from the yellow area.

22.5 Color Recognition of Plant Images

Through the established color database, the plant image is processed, and finally, the corresponding color result is obtained. First, we import the image of the plant into the color analysis system, and then, we process the image through the color analysis system. In the color analysis system, we want to remove the background of the image. This study uses the Grab cut object matting algorithm. After removing the background of the target, we frame the target. After determining the scope of the target, the system will automatically analyze the color of the target and then give an accurate result. This study is using images of three petals, and the color analysis system will identify the colors of the three petals. Figure 22.4 is the analysis of plant images.

In this study, the color recognition system is used to process the plants, and then, the color characteristics of the plants can be obtained. The results are mainly divided into three categories as follows: (1) color distribution, (2) Munsell color cards, and

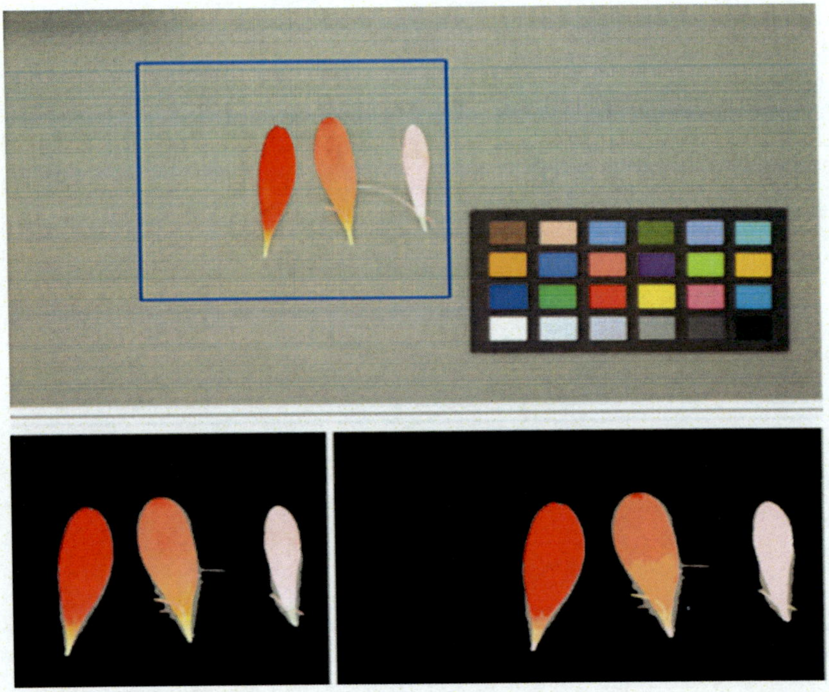

Fig. 22.4 Plant image analysis

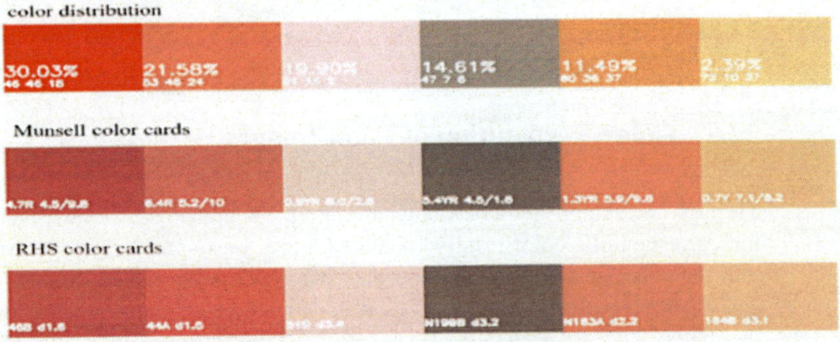

Fig. 22.5 Color analysis results

(3) RHS color cards. This will help us to accurately judge the color characteristics of plants. It can help botanical scholars to quickly and qualitatively analyze the color of plants. The color analysis system can also give some quantitative data, such as the value of CIE $L^*a^*b^*$ and the value of RGB. This helps scholars to analyze the color of plants through data. We will show the results after processing with the color analysis system. Figure 22.5 shows the relevant processing results.

Through the image processing results, we can judge that the plant image contains six color systems. Among them, the main color is dark red, accounting for 30.03%, and the secondary color is medium red, accounting for 21.58%. We can see the color distribution from the picture, the results of the Munsell color card and the RHS color card. Through this color analysis system, we can qualitatively analyze the color of plants and quantitatively analyze the color categories of plants.

22.6 Conclusions

This paper mainly identifies the color features of plant images. We obtain the standard color comparison card sample set and divide the standard color comparison card sample set into a training set and a test set. A sample set of standard color charts contains (1) the RGB value of the color number and (2) CIE $L^*a^*b^*$ value of color number. Then, the recognition accuracy test is carried out by using the recognition model of the test set, and based on the result of the recognition accuracy test, a target color database is established. This study tested the RHS color card through the color recognition system. We tested a total of 884 color numbers. In the test, 860 color numbers were identified correctly, and 24 color numbers were incorrectly identified. According to the number of correct color numbers, we can judge the accuracy of the system. This color recognition method can avoid the error of manual visual comparison. Its advantages are as follows: (1) It is relatively stable and not easily affected by environmental factors. (2) The recognition efficiency is high, and the recognition accuracy can reach 97%.

References

1. Bramão, I., Reis, A., Petersson, K.M.: The role of color information on object recognition: a review and meta-analysis. Acta Physiol. (Oxf) **138**(1), 244–253 (2011)
2. Van De Sande, K., Gevers, T., Snoek, C.: Evaluating color descriptors for object and scene recognition. IEEE Trans. Pattern Anal. Mach. Intell. 1582–1596 (2010)
3. John, E.T., Arthur, C.H.: An analysis of the original Munsell color system. Opt. Soc. Am. **30**, 587–590 (1940)
4. Voss, D.H.: The royal horticultural society color chart 2001 (2002)
5. Gevers, T., Smeulders, A.W.M.: Color-based object recognition. Pattern Recogn. **32**(3), 453–464 (1999)
6. Rachmadi, R.F., Purnama, I.: Vehicle color recognition using convolutional neural network. arXiv:1510.07391 (2015)
7. Kumar, T., Verma, K.: A theory based on conversion of RGB image to gray image. Int. J. Comput. Appl. **7**(2), 7–10 (2010)
8. Zhang, X., Wandell, B.A.: A spatial extension of CIELAB for digital color-image reproduction. J. Soc. Inform. Displ. **5**(1), 61–63 (1997)
9. Cochrane, S.: The Munsell color system: a scientific compromise from the world of art. Stud. Hist. Philos. Sci. Part A **47**, 26–41 (2014)

10. Parkkinen, J.P.S., Hallikainen, J., Jaaskelainen, T.: Characteristic spectra of Munsell colors. JOSA A. **6**(2), 318–322 (1989)
11. Burns, S.A., Cohen, J.B., Kuznetsov, E.N.: The Munsell color system in fundamental color space. Color. Res. Appl. **15**(1), 29–51 (1990)
12. Microsoft Homepage. https://visualstudio.microsoft.com/zh-hans/vs. Accessed 20 Dec 2022

Chapter 23
Research Status of Underwater Fishing Equipment Technology

Fulu Ji and Quanliang Liu

Abstract This paper mainly introduces the research status of related technologies of marine fishing equipment at home and abroad, including underwater image recognition processing, kinematic model parameter identification and research and development of motion control system and other related technologies. Some problems existing in marine operation fishing equipment are summarized, and some prospects for the future development trend of marine operation fishing equipment are put forward on this basis.

23.1 Preface

The ocean is the cradle of life on earth, and it is also the last area on earth that human beings can develop and utilize. Marine resources are very rich, and marine aquatic products provide human beings with high-quality protein, especially marine fish and aquatic products. However, in the recent years, under the situation of long-term high-intensity fishing pressure and the deteriorating marine ecological environment, it is an indisputable fact that resource and environmental constraints have intensified and fishery resources have continued to decline. According to the fishing regulations issued by the state, the fish catch is lower than the growth of fishery resources, determines the total catchable amount of fishery resources, implements the catch quota system and requires the upgrading and updating of fishing equipment. Looking at relevant domestic and foreign kinds of literature, it is found that underwater robots and related underwater intelligent equipment are widely used in the aquatic fishing and aquaculture industries for detection and protection. The research and application of underwater intelligent fishery equipment in China is still in its infancy, and the corresponding research is less. Therefore, the development of marine fishery underwater intelligent equipment in China fills the gap in the research field of domestic

F. Ji · Q. Liu (✉)
Zhejiang Ocean University, Zjou, Zhoushan, China
e-mail: 1540832871@qq.com

S. Patnaik et al. (eds.), *3D Imaging—Multidimensional Signal Processing and Deep Learning*, Smart Innovation, Systems and Technologies 348,
https://doi.org/10.1007/978-981-99-1145-5_23

fishery underwater intelligent equipment and helps the vigorous development of the deep-blue fishing industry.

23.2 Research Status of Underwater Image Recognition

There are a lot of tiny particles in seawater. Because the existence of these particles will affect the uniform distribution of the light beam into the water and because water has the effect of absorbing and scattering light, it will cause low contrast and blurred underwater images, color attenuation, noise, etc. In order to improve the quality of the images captured in the water and improve the problems in the process of image recognition, it is necessary to preprocess the underwater images. At present, the commonly used underwater image preprocessing includes image enhancement and segmentation [1]. Image enhancement mainly enhances the overall or local features of an image, while image segmentation combines the main information with the secondary information to retain the useful part of the image. Make a distinction, keep the useful parts of the image and remove the useless parts.

In the field of underwater image enhancement, experts and scholars at home and abroad have conducted research. For example, LIU Y and CHAN W proposed the use of the spatial domain method to enhance images [2]. This algorithm uses fuzzy logic rules to determine color parameters to minimize the color temperature difference of various light sources. This algorithm has been implemented in the newly designed electronic static camera and has led to a high improvement in image quality; Daniel L. Bongiorno et al. illuminated natural light and artificial light with object distances of 1.4, 1.55 and 1.77 m, respectively, through the laboratory pool, colorimetric plate and spectrometer. On this basis, they studied the qualitative evaluation of visual effect by color correction model [3]; Muhammad Suzuri Hitam et al. used the restricted contrast adaptive histogram equalization and Rayleigh distribution application and image enhancement through the marine vegetation of Terengganu Islands, Malaysia, and made a quantitative evaluation of the covariance and signal-to-noise ratio of the captured images [4]; Tian Yonghong et al. deduced a new underwater dark channel to estimate the scattering rate and proposed an effective method to estimate the background light in the underwater optical model. The experimental results show that the algorithm can process underwater images well, especially deep-sea images and images captured from turbid waters [5]; Xin Luan et al. corrected non-uniform lighting by homomorphic filtering, equalized the RGB color space histogram of the image and denoised it by wavelet transform [6]. The contrast and clarity of the image after processing with this algorithm are better; Yao Kai uses a wavelet definition to enhance image fusion [7]. This method can overcome the impact of a complex optical environment in water and fuse multi-focus images to obtain a good fusion image. However, it fails to obtain the desired processing effect for low contrast, blurring and other problems of underwater images; Fei Lei et al. proposed an adaptive wavelet transform denoising method to enhance underwater images. Through testing, it was found that this method can improve image quality and enhance key point matching and edge detection [8].

In the field of underwater image segmentation, the same domestic and foreign experts and scholars have also conducted research. For example, Li Tao and others have studied the fuzzy entropy segmentation algorithm, and on this basis, they have also made a new definition, using the POS algorithm to search and segment the threshold [9]; Mehdi Fatan et al. proposed a texture-based underwater cable recognition and segmentation method, which has been used in multiple scenes with high recognition accuracy [10]; Ronald Philipo and Christian Barat studied a fully automatic active contour detection method for segmentation of underwater fixed objects. Through comparative experiments, it was found that this method has strong adaptability and high segmentation accuracy [11]; Donghoon Kim et al. proposed to segment the target path manually set underwater based on color features and used the correlation coefficient method to evaluate the weight value of each feature of the target. This algorithm can achieve multi-target recognition without being affected by the size of the target object [12]; Chen Zhe et al. proposed an underwater target segmentation algorithm based on the weighted fusion of multi-feature information; adaptively estimated the weights of light intensity, color and texture direction features based on machine learning method; further modified the fusion features by region growth algorithm; and improved the Lu Chun property W of the feature model to adapt to underwater environments with different optical characteristics [13]; Dario Lodi Rizzini et al. also studied the target detection algorithm based on multiple features and carried out automatic detection experiments on plastic pipes manually placed in natural waters. Compared with traditional methods, they have stronger adaptability to the environment [14].

Underwater image recognition research has been applied to the field of underwater robot vision and has been widely used in underwater exploration and underwater task execution, greatly improving the work efficiency of underwater robots and providing valuable resources for human exploration of the underwater world.

23.3 Research Status of Kinematic Parameter Identification

As we all know, the marine environment has a lot of uncertainties and instability, especially the ocean currents in the ocean. If the marine fishing equipment is to be able to operate stably underwater, it must have a certain anti-interference performance. Therefore, an accurate kinematics model is very important for marine fishing equipment, and the identification of kinematics parameters is the basis for calculating the kinematics model. According to relevant literature, there are three main types of kinematic modeling for offshore operation equipment: a mathematical model based on Newton Euler equation, a mathematical model based on linear system theory and a mathematical model based on the neural network [15]. The kinematics mathematical model proposed above has strong coupling and nonlinearity, and its parameters need to be identified according to the actual operation situation. At present, the identification of the kinematic model mainly includes the following methods: computational fluid dynamics method, full-scale or reduced-scale drag test method, empirical formula approximate estimation method and model identification method [16].

The computational fluid dynamics method uses computational fluid dynamics through computer software, uses finite discrete points to replace the continuous pressure field and velocity field in the time–space domain, establishes the algebraic relationship between discrete points through a certain relationship logic, and then simulates the underwater robot's motion in the water, solves the approximate value of field variables and represents the hydrodynamic parameters approximately through logical conversion. Chin uses ANSYS software and uses the SSTk-ε turbulence model to numerically estimate the ROV water damping coefficient. The experimental results are highly consistent with the physical experimental results of the pool [17]. Eng used the CFD method to simulate and analyze the sway, surge, heave and yaw motions of ROV and obtained that the error between the added mass and the actual value is 2% [18]; Xu Mengmeng et al. used the overlapping grid method to identify the parameters of the rotational kinematics model of the complex structure ROV in STAR-CCM+, and then carried out numerical simulation and laboratory research, respectively, with very similar results [19]; Yu Huanan obtained some parameters of the hydrodynamic model by using the least squares method. At the same time, he also found that the parameter estimation when using the least squares method to identify the model parameters is uniformly convergent and unbiased when the noise is zero mean and white noise [20]; Nahon and Chen have made contributions to the calculation of submarines and the calculation of new models by using empirical formula method. However, this method is only applicable to underwater robots with regular geometric shapes and slender fluid structures and does not apply to underwater robots with complex shapes and open-rack structures [21].

Due to the strong disturbance of the underwater environment, the control system of underwater fishing equipment is required to have good robustness. The key to control system modeling is the identification of its parameters. Through the research of many scholars, it has laid a good foundation for obtaining the parameters of the mathematical model of underwater operation equipment control systems.

23.4 Current Status of Motion Control Research

The uncertainty and instability of the marine environment will lead to the existence of unknown nonlinear hydrodynamic effects and parameter uncertainty of marine operation equipment. The control system design problems of equipment include various nonlinearity and modeling uncertainty, including fluid dynamics nonlinearity, inertia nonlinearity and coupling problems with DOF. Although there are many methods to model the motion control model, most of them are insufficient, so the uncertainty is still high [22]. According to this situation, experts and scholars at home and abroad have carried out a lot of research and discussion on the marine operation equipment control system and have drawn some constructive conclusions.

Norwegian scholar Fossen has carried out a series of research and summary on the dynamic model, navigation, positioning and control methods of ships and underwater vehicles; derived in detail the dynamic equations in the form of components and matrices, frequency domain, time domain and different motion coordinate systems; and sorted out and analyzed various dynamic models used in the control design of ocean vehicles. Without considering the interaction between the underwater robot and water, according to the Newton Lagrangian mechanical equation of the rigid body in the fluid, the six-degree-of-freedom nonlinear motion equation of the underwater robot in the moving coordinate system can be described as [23]:

$$M\dot{v} + C(v)v + D(v)v + g(\eta) = \tau_{\mathrm{E}} + \tau \tag{23.1}$$

where M is the inertia matrix of ROV; $C(v)$ is the Coriolis force and centrifugal force matrix; $D(v)$ is the hydrodynamic damping matrix, a vector consisting of potential damping induced by radiation, linear surface friction, wave drift damping and vortex shedding damping; $g(\eta)$ is the recovery vector formed by ROV buoyancy and gravity term; $v = [u, v, w, p, q, r]^T$ is the linear velocity and angular velocity vector of ROV in the moving coordinate system; τ_E is the environmental force and moment vector caused by current and wave; and τ is the vector composed of thrust and torque formed by ROV propeller. Compared with the model proposed by other experts and scholars, the matrix form and dynamic model proposed by Fossen are simple, practical and convenient for analysis and design, so the model has been used up to now [24].

The motion control of the underwater vehicle (ROV) depends on its motion mathematical model, and the moving coordinate system is the foundation of the ROV motion mathematical model. This paper uses the coordinate system [25] recommended by the terminology report of the Society of Shipbuilding and Marine Engineering and the International Pool Conference. The specific form of the coordinate system is shown in Fig. 23.1.

The underwater vehicle has six degrees of freedom in the moving coordinate system, and it moves along the XYZ axis in a straight line: surge (u), sway (v) and heave (w); rotating motion around the XYZ axis: roll (p), pitch (q) and yaw (r).

For the expansion of the six-degree-of-freedom mathematical model of ROV motion in the moving system, it can be studied from the two points of translation and rotation. Before the analysis, it is necessary to set the gravity center coordinate $r_G(x_G, y_G, z_G)$ and buoyancy center coordinate $r_B(x_B, y_B, z_B)$ of the ROV. Set the inertia matrix that the gravity center of the ROV does not coincide with the origin of the power system as I, and the specific expression of I is:

$$I = \begin{pmatrix} I_{xx} & -I_{xy} & -I_{xz} \\ -I_{yx} & I_{yy} & -I_{yz} \\ -I_{zx} & -I_{zy} & I_{zz} \end{pmatrix} \tag{23.2}$$

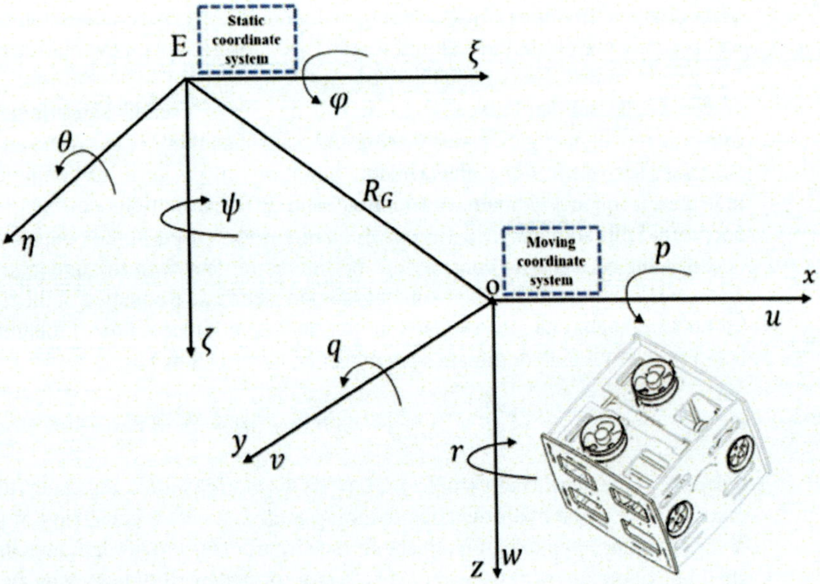

Fig. 23.1 ROV space motion coordinate system

According to the rigid body momentum theorem and Formula (23.1), the linear motion expression of ROV on the three coordinate axes of the moving system is:

$$\begin{cases} m\left[\dot{u} - vr + wq - x_G\left(q^2 + r^2\right) + y_G(pq - \dot{r}) + z_G(pr + \dot{q})\right] = X \\ m\left[\dot{v} - wp + ur - y_G\left(r^2 + p^2\right) + z_G(qr - \dot{p}) + x_G(qp + \dot{r})\right] = Y \\ m\left[\dot{w} - uq + vp - z_G\left(p^2 + q^2\right) + x_G(rp - \dot{q}) + y_G(rq + \dot{p})\right] = Z \end{cases} \tag{23.3}$$

Similarly, according to the theorem of the moment of momentum and Eqs. (23.1) and (23.2), the rotational motion expression of ROV on the three coordinate axes of the moving system is:

$$\begin{cases} I_{xx}\dot{p} + I_{xy}\dot{q} + I_{xz}\dot{r} + \left(I_{zx}p + I_{zy}q + I_{zz}r\right)q - \left(I_{yx}p + I_{yy}q + I_{yz}r\right)r \\ +m[y_G(\dot{w} + vp - uq) - z_G(\dot{v} + ur - wp)] = K \\ I_{yx}\dot{p} + I_{yy}\dot{q} + I_{yz}\dot{r} + \left(I_{xx}p + I_{xy}q + I_{xz}r\right)r - \left(I_{zx}p + I_{zy}q + I_{zz}r\right)p \\ +m[z_G(\dot{u} + wp - ur) - x_G(\dot{w} + vp - uq)] = M \\ I_{zx}\dot{p} + I_{zy}\dot{q} + I_{zz}\dot{r} + \left(I_{yx}p + I_{yy}q + I_{yz}r\right)p - \left(I_{xx}p + I_{xy}q + I_{xz}r\right)q \\ +m[x_G(\dot{v} + ur - wp) - y_G(\dot{u} + wq - vr)] = N \end{cases} \tag{23.4}$$

where X is ROV surge hydrodynamic force, Y is ROV sway hydrodynamic force, Z is ROV heave hydrodynamic force, K is ROV roll hydrodynamic moment, M is ROV pitch hydrodynamic moment, and N is ROV yaw hydrodynamic moment.

PID control is a kind of linear control, which is simple to control, easy to realize and does not need an accurate mathematical model. It is the most mature control method used in the industrial control field at present. However, due to the strong coupling, nonlinearity and time variation of marine operation equipment, PID control cannot achieve ideal results. In the recent years, modern intelligent control technology has been widely used in the control system of marine operation equipment and has achieved good control results. Modern intelligent control has a better control effect because it can more effectively use the uncertainty of fluid dynamics and shows excellent anti-interference ability. In addition, it does not require accurate models of equipment and environment, thus significantly reducing the complexity of the design. Modern intelligent control includes neural network control (NN), sliding mode control (SLC), fuzzy logic control (FLC), adaptive control (APC), cascade PID control and the combination of modern intelligent control and PID control. Silvia M. Zanoli et al. realized the simulation and test of depth control and fixed-point control of the underwater vehicle using PID and achieved good control performance [26]. Khodayari used an adaptive control algorithm and fuzzy control algorithm to jointly adjust and predict the parameters of the PID control system, effectively solving the problem of PID real-time parameter tuning [27]. Yu Jiancheng et al. applied a generalized dynamic fuzzy neural network to implement direct adaptive control of ROV and used the Lyapunov stability theory to prove that the method is effective [28]. Zhou Huan yin et al. decomposed the control system into a nonlinear uncertain part and an approximate linear part and used a neural network algorithm and dynamic feedback control algorithm to control the two parts, respectively, which proved that this method has strong dynamic performance and robustness [29]. Iranian scholars J. Javadi Moghaddam and A Bagheri designed an adaptive neuro-fuzzy sliding mode control system (ANFSGA) based on a genetic algorithm for ROV equipped with four thrusters. Through testing and research, it is found that the control system has good control performance for R0V [30]. Christian Mai et al. tested and simulated the control system of the industrial ROV VideoRay 4 PRO, designed a linear quadratic optimal controller with full state feedback and simulated the control system with a nonlinear model. The test and simulation results show that the control system has good robustness [31]. Enrico Anderlini of the University of London, Giles Thomas and Gordon G. Parker of the University of Science and Technology of Michigan have studied the trajectory control of ROV carrying heavy objects and designed an adaptive model predictive control strategy (AMPC), which will help to select an appropriate control scheme for future unmanned aerial vehicles that independently perform maintenance tasks [32]. Wang Jianguo used PID control and fuzzy control for reference to design an s-plane controller for the underwater vehicle and carried out verification and simulation calculations. The results show that the s-plane controller can improve the control performance of the underwater vehicle very well [33]. Fu Yuxin carried out the simulation test of the track-tracking control of the unmanned underwater vehicle based on the fuzzy neural network. The simulation test results show that the control method can be well applied to the motion control of the unmanned underwater vehicle [34]. Zhu Qi designed the PID controller and quasi-sliding mode controller of the UVMS system, carried out motion control simulation for the propulsion configuration and fault state of the eight propellers and six propellers, respectively, and carried out simulation tests for the movement of the manipulator in hover state, in still water state and flat flow environment object grasping [35]. The communication and state control system of the robot developed by Zhu Daqi and others using the control algorithm has been tested to show that the control system can run stably to complete the task. Song Xin and others added an intelligent closed-loop control algorithm of the computer to the ultra-small underwater vehicle with the open-loop control system so that ROV is controlled by both human and machine intelligence. Experiments show that this strategy greatly enhances the adaptability and robustness of the control system [36]. Wang Jianhua and others analyzed the force of ROV in the complex underwater environment, established the kinematic mathematical model of ROV pitch angle stability control and therefore designed a set of cascade controllers [37]. Zeng Dewei designed an adaptive control law based on deterministic learning and feedback linearization and simulated the horizontal, vertical and three-dimensional trajectory tracking of the underactuated AUV. The results show that the designed controller is effective and achieves a satisfactory control effect [38].

23.5 Summary

With the development of science and technology, marine operation equipment technology is becoming more and more advanced. In the past, it can complete the functions of detection and monitoring; to now, it can complete the work of catching relatively static marine products. For the current underwater image recognition problem of marine fishing equipment, airspace method, restricted contrast adaptive histogram equalization, underwater dark signal channel and homomorphic filtering method can be used to deal with it, and the above methods can all show good performance; at present, computational fluid dynamics method, full-scale or scaled drug test method, empirical formula approximate estimation method and model identification method can be used to identify the kinematic parameters of marine fishing equipment, and an accurate parameter is often helpful for studying the modeling of underwater equipment control system; for the current motion control method of marine fishing equipment, more modern methods can be adopted, such as intelligent control including neural network control (NN), sliding mode control (SLC), fuzzy logic control (FLC), adaptive control (APC), cascade PID control and the control method combining modern intelligent control with PID control. In the future, marine fishing equipment will be more capable of artificial intelligence. The three issues studied above are the key starting point. Therefore, the content described in this paper will have a certain guiding significance for the research of underwater fishing equipment technology.

References

1. Schettini, R., Corchs, S.: Underwater image processing: state of the art of restoration and image enhancement methods. EURASIP J. Adv. Signal Process. **2010**, 1–14 (2010)
2. Liu, Y.C., Chan, W.H., Chen, Y.Q.: Automatic white balance for digital still camera. IEEE Trans. Consum. Electron. **41**(3), 460–466 (1995)
3. Bongiorno, D.L., Bryson, M., Williams, S.B.: Dynamic spectral-based underwater colour correction. In: 2013 MTS/IEEE OCEANS-Bergen, pp. 1–9. IEEE (2013)
4. Hitam, M.S., Awalludin, E.A., Yussof, W.N.J.H.W., Bachok, Z.: Mixture contrast limited adaptive histogram equalization for underwater image enhancement. In: 2013 International Conference on Computer Applications Technology (ICCAT), pp. 1–5. IEEE (2013)
5. Wen, H.C., Tian, Y.H., Huang, T.J., et al. Single underwater image enhancement with a new optical model. In: 2013 IEEE International Symposium on Circuits and Systems (ISCAS), pp. 753–756. IEEE (2013)
6. Luan, X., Hou, G., Sun, Z., Wang, Y., Song, D., Wang, S.: Underwater color image enhancement using combining schemes. Mar. Technol. Soc. J. **48**(3) (2014)
7. Yao, K.: Research on underwater image fusion enhancement based on wavelet definition calculation. Electron. Meas. Technol. **38**(2), 64–67 (2015)
8. Lei, F., Wang, Y.Y.: The research of underwater image de-noising method based on adaptive wavelet transform. In: The 26th Chinese Control and Decision Conference (2014 CCDC), pp. 2521–2525. IEEE (2014)
9. Li, T., Tang, X.D., Pang, Y.J.: Underwater image segmentation based on improved PSO and fuzzy entropy. Ocean Eng. **28**, 128–133 (2010)
10. Fatan, M., Daliri, M.R., Shahri, A.M.: Underwater cable detection in the images using edge classification based on texture information. Measurement **91**, 309–317 (2016)
11. Barat, C., Phlypo, R.: A fully automated method to detect and segment a manufactured object in an underwater color image. EURASIP J. Adv. Signal Process. **2010**, 1–10 (2010)
12. Kim, D., Lee, D., Myung, H., Choi, H.T.: Artificial landmark-based underwater localization for AUVs using weighted template matching. Intel. Serv. Robot. **7**(3), 175–184 (2014)
13. Zhe, C., Wang, H., Xu, L., et al.: Visual-adaptation-mechanism based underwater object extraction. Opt. Laser Technol. **56**(1), 119–130 (2014)
14. Lee, D., Kim, G., Kim, D., et al.: Vision-based object detection and tracking for autonomous navigation of underwater robots. Ocean Eng. **48**, 59–68 (2012)
15. Fossen, T.I.: Guidance and control of ocean vehicles. Doctors Thesis, University of Trondheim, Norway, Printed By John Wiley & Sons, Chichester, England (1999). ISBN: 0 471 94113 1
16. Deng, Z.G., Zhu, D.Q., Fang, J.A.: A review of parameter identification methods for underwater robot dynamics model. J. Shanghai Marit. Univ. **35**(2), 74–80 (2014)
17. Chin, C., Lau, M.: Modeling and testing of hydrodynamic damping model for a complex-shaped remotely-operated vehicle for control. J. Mar. Sci. Appl. **11**(2), 150–163 (2012)
18. Eng, Y.H., Chin, C.S., Lau, M.W.S.: Added mass computation for control of an open-frame remotely-operated vehicle: application using WAMIT and MATLAB. J. Mar. Sci. Technol. **22**(4), 405–416 (2014)
19. Xu, M.M.: Dynamic modeling and control of ROV with complex shape. Shanghai Jiaotong University (2017)
20. Yu, H.N.: Research on identification and control technology of open-frame underwater robot. Harbin Engineering University (2003)
21. Nahon, M.: A simplified dynamics model for autonomous underwater vehicles. In: Proceedings of Symposium on Autonomous Underwater Vehicle Technology, pp. 373–379. IEEE (1996)
22. Chen, Y.: Modular modeling and control for autonomous underwater vehicle (AUV) (2008)
23. Fossen, T.I.: Marine Control Systems (2002)
24. Fossen, T.I.: Handbook of Marine Craft Hydrodynamics and Motion Control. Wiley (2011)
25. Jiang, X.S., Feng, X.S., Wang, D.T.: Underwater Robot: Liaoning Science and Technology Press (2000)

26. Zanoli, S.M., Conte, G.: Remotely operated vehicle depth control. Control. Eng. Pract. **11**(4), 453–459 (2003)
27. Khodayari, M.H., Balochian, S.: Modeling and control of autonomous underwater vehicle (AUV) in heading and depth attitude via self-adaptive fuzzy PID controller. J. Mar. Sci. Technol. **20**(3), 559–578 (2015)
28. Yu, J.C., Zhang, A.Q., Wang, X.H., et al.: Direct adaptive control of underwater robot based on fuzzy neural network. J. Autom. **8**, 840–846 (2007)
29. Zhou, H.Y., Liu, K.Z., Feng, X.S.: Dynamic feedback control of autonomous underwater robot based on neural networks. J. Motors Control **15**(7), 87–93 (2011)
30. Javadi-Moghaddam, J., Bagheri, A.: An adaptive neuro-fuzzy sliding mode based genetic algorithm control system for under water remotely operated vehicle. Expert Syst. Appl. **37**(1), 647–660 (2010)
31. Mai, C., Pedersen, S., Hansen, L., Jepsen, K., Yang, Z.: Modeling and control of industrial ROV's for semi-autonomous subsea maintenance services. IFAC-PapersOnLine **50**(1), 13686–13691 (2017)
32. Anderlini, E., Parker, G.G., Thomas, G.: Control of a ROV carrying an object. Ocean Eng. **165**, 307–318 (2018)
33. Wang, J.G.: Research on motion control and fault diagnosis technology of underwater robot. Harbin Engineering University (2011)
34. Fu, Y.X.: Track tracking control of unmanned underwater vehicle based on fuzzy neural network. Dalian Maritime University, Dalian (2017)
35. Zhu, Q.: Research on attitude control method of operational underwater robot. Zhejiang University (2018)
36. Song, X., Ye, J.W., Liang, F.L., et al.: Improved design and intelligent control system of subminiature underwater vehicle. Robotics **6**, 596–600 (2007)
37. Wang, J.H., Song, Y., Wei, G.L., et al.: Application of cascade PID control in pitch control system of underwater vehicle. J. Shanghai Univ. Technol. **39**(3), 229–235 (2017)
38. Zeng, D.W.: Research on trajectory tracking control method of underwater vehicle. South China University of Technology (2017)

Chapter 24
Research on Network Traffic Classification Method Based on CNN–RNN

Zhaotao Wu and Zhaohua Long

Abstract With the continued development of the information society, the scale of the Internet grows ever larger, and the enormous amount of traffic data throughput in the Internet is so diverse that the difficulty of the network traffic classification problem is constantly challenged, and the speed and accuracy of the classifier are put forward to higher standards. It is critical to understand how to continuously optimize network traffic classification techniques using existing technologies for implementing network security censorship, strengthening network security management, detecting and defending against network intrusion, and so on. The traditional port-based classification method is no longer reliable in today's increasingly complex network environment, and the deep packet inspection technique is inapplicable to the classification of encrypted network traffic. The machine learning-based classification method has good classification performance and solves the problem that deep packet inspection cannot identify encrypted network traffic, which has been a hot research topic in network traffic classification in the recent years. To address the shortcomings of existing methods, this paper proposes a classification method combining recurrent neural network and convolutional neural network within the framework of machine learning.

24.1 Introduction

As the Internet continues to expand, new types of applications emerge. With the proliferation of applications comes an increase in network traffic, and different types of traffic have different characteristics. The goal of traffic classification is to identify traffic classes based on distinguishing features. The classification of network traffic is required for operators for several reasons: on the one hand, from the perspective of user quality of service (QoS), traffic classification is the first step to ensuring

Z. Wu (✉) · Z. Long
School of Computer Science and Technology, Chongqing University of Posts and Telecommunications, Chongqing 400065, China
e-mail: 1715633164@qq.com

S. Patnaik et al. (eds.), *3D Imaging—Multidimensional Signal Processing and Deep Learning*, Smart Innovation, Systems and Technologies 348,
https://doi.org/10.1007/978-981-99-1145-5_24

QoS, which is a prerequisite for providing differentiated services based on the needs of different service types; on the other hand, from the perspective of security, traffic classification is the first step to detecting abnormal network traffic, which can better protect networks. With increasing user demand for privacy protection and the continuous development of encryption technology, more and more traffic has been encrypted and processed in the recent years, posing new challenges to network traffic classification.

Traditional network traffic classification methods are divided into two categories: early network traffic identification methods were based on port numbers [1], i.e., identification based on the protocol number corresponding to the port number; however, with the advent of port obfuscation technology, port number-based traffic identification methods are no longer reliable, and the accuracy rate gradually decreases. The network traffic identification and classification technique based on deep packet inspection (DPI) [2], also known as load-based traffic classification technique, falls into another category. By pattern matching the feature information in the packet with the feature signature of known applications, the DPI technique identifies the usage type of data traffic. The obvious disadvantage of DPI technology is that it cannot recognize encrypted message loads, and as encrypted messages appear in more and more applications, DPI technology's limitations become more apparent.

Researchers have begun to investigate new traffic classification methods as traditional traffic classification methods continue to fail. Researchers have paid close attention to machine learning techniques that have emerged in the recent years. Machine learning techniques are more automated and intelligent than traditional classification methods and can classify traffic based on statistical characteristics, avoiding the impact of traffic encryption. In light of this benefit, researchers have proposed methods for traffic classification based on widely used machine learning algorithms, such as the decision tree algorithm, random forest algorithm, and support vector machine algorithm. All of these classification methods have high accuracy and are widely used in academia and industry. However, machine learning-based traffic classification methods necessitate expert knowledge to extract and filter traffic features, which can consume a significant amount of human resources. Taking this into account, researchers have proposed a new end-to-end traffic classification method based on deep learning [3]. The deep learning-based approach can directly classify traffic based on raw traffic data without human feature extraction, saving manpower on the one hand while ensuring classification accuracy on the other, making it a research focus in academia.

In the recent years, deep learning-based traffic classification methods have emerged. Deep learning is based on the neural network concept in machine learning, but it eliminates the need to manually set up features, has strong feature learning capability due to multiple hidden layers, can automatically extract features from the dataset, and is continuously optimized to ensure classification accuracy. The following is a summary of existing methods for classifying network traffic.

24.2 Network Traffic Classification Method

24.2.1 Machine Learning-Based Network Traffic Classification

There have been numerous machine learning-based traffic classification methods proposed. Auld et al. [4] proposed a Bayesian-based neural network that has been trained to recognize well-known P2P protocols such as Kazza, BitTorrent, and GnuTella. This trained neural network, however, has a half-life, so classification accuracy decreases over time as the composition of Internet traffic changes. Moore et al. [5] classify streams based on statistical features such as packet size, mean of packet arrival interval, variance, and other features, with a final classification accuracy of 96% using a simple Bayesian classifier and a kernel density estimator. Draper et al. [6] used KNN and the C4.5 decision tree algorithm to characterize network traffic based on time-dependent features such as maximum–minimum packet arrival intervals, with a final classification recall of 92%. On the VPN dataset, the C4.5 algorithm achieved a recall of around 88%. Shen et al. [7] proposed a decentralized application identification method in 2019, proposing feature fusion using kernel functions based on statistical features of bidirectional data streams, followed by further feature filtering to achieve 92% classification accuracy.

Feature extraction is a critical aspect of machine learning tasks, as it involves creating a batch of abstract data that reflects the characteristics of the original data. When classifying network traffic using deep stream detection, common stream features include average packet size, stream duration, total number of packets, packets per second, stream bandwidth, and so on. In the traditional machine learning domain, these features are defined based on data characteristics and heavily rely on the person defining the features' industry experience [8]. Choosing a good feature is frequently a difficult task, necessitating a high level of expertise and industry experience on the part of the feature engineers, and the final model performance is subject to great uncertainty.

24.2.2 Deep Learning-Based Network Traffic Classification

Deep learning-based methods can learn features autonomously rather than selecting them based on expert experience. Deep learning is a highly desirable method for traffic classification because of this feature. Furthermore, deep learning models are end-to-end models that can learn the nonlinear relationship between the original input and the corresponding output without the need to divide the problem into two parts: feature selection and classifier classification.

With the advancement of deep learning technology, an increasing number of researchers have applied deep learning to the study of network traffic classification. Moreira et al. [9] used deep convolutional neural networks to classify network traffic.

They encoded the packet header information and load data to generate the packet view vector, and then fed the view vector into the convolutional neural network for type recognition. The model achieved good recognition accuracy after training on a large number of samples. Wang et al. [3] proposed a one-dimensional convolutional neural network-based encrypted flow classification method that uses CNN as the learning algorithm to automatically learn features from unprocessed traffic and then uses the high-level features as input to the softmax layer, which then outputs the predicted application classification labels. This method avoids the steps of traditional machine learning methods such as feature design, feature extraction, and feature selection and produces better classification results. However, its performance is heavily dependent on the training dataset, and the larger the dataset, the higher the accuracy.

Deep learning-based methods can learn features autonomously rather than selecting them based on expert experience [10]. Deep learning is a highly desirable method for traffic classification because of this feature. Furthermore, deep learning models are end-to-end models that can learn the nonlinear relationship directly. Between deep learning methods based on packet raw byte features, as well as deep learning methods based on in-stream packet sequence features, each has advantages and disadvantages. The deep learning method based on packet raw byte features has the advantage of being able to perform inference directly based on the packet byte content and achieve real-time classification, but the classification result is dependent on the content of the packet load. When traffic is encrypted, a portion of the packet load is unavailable; additionally, the IP address and port fields of packets can have a significant impact on the classification effect and cause overfitting problems. The deep learning method based on packet sequence features within the flow has the advantage of not relying on the original content of the packet and is more flexible for encrypted traffic, but it must wait for the packets to form a temporal sequence, so the classification is less real time.

24.3 Related Work

24.3.1 Datasets and Preprocessing

Reliable public datasets serve as the foundation for traffic classification research. Studies based on non-public datasets have a negative impact on the credibility of the results and make it difficult for other researchers to replicate the experimental findings. The PCAP dataset used in this paper is one of the most commonly used encrypted traffic datasets in network traffic classification studies. During two weeks in 2016, the dataset was gathered by crawling the most visited HTTPS websites on Google Chrome and Mozilla Firefox twice a day.

We perform various preprocessing on the pcap file to generate the training data. Because the pcap file could contain a variety of traffic from the local machine, we used an SSL filter in Wireshark to capture only HTTPS traffic.

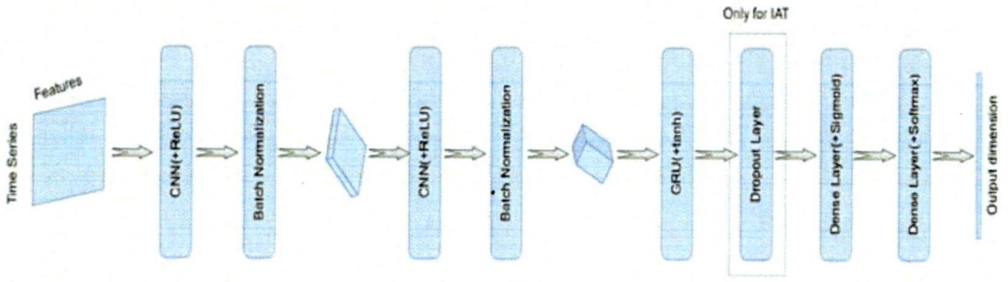

Fig. 24.1 Basic model for network traffic classification

24.3.2 *Basic Model*

For sequence feature classification, the deep learning architecture used in this paper primarily employs convolutional neural networks (CNN) and gated recurrent units (GRU). Figure 24.1 depicts the basic model framework.

24.3.3 *Experimental Environment*

The main configuration of the physical machine is CPU Intel i7-8700 K, GPU NVIDIA GeForce GTX1080Ti, and memory 64G. The operating system is 64-bit Windows 10, and the deep learning framework used is PyTorch.

24.4 Experimental Process

The basic model developed in this paper begins with a CNN trained only on packet sequences, followed by a structure based on Keras' GRU, and finally a layer of softmax fully connected dense layers. The number of neurons in the output matches the number of SNI classes in the dataset. During the ten periods of training, Adam was the primary optimizer, and the loss algorithm was sparse classification cross-entropy.

To begin, we performed a tenfold cross-validation accuracy analysis of the different algorithms in the first day's sampled data collocation set, and the results are shown in Fig. 24.2, where the random forest classifier maintains a classification accuracy of more than 85% for a minimum number of connections greater than 30. Figure 24.3 also shows the accuracy analysis for the Auto-Sklearn algorithm. The accuracy of the Auto-Sklearn correlation algorithm is slightly lower than that of the original random forest classifier, as shown in the figure, owing to the fact that it selects fewer data estimates. However, it outperforms the RNN classifier that uses the packet size sequence as the baseline feature.

After reviewing the preliminary results, we discovered that two major issues must be addressed in order to improve the accuracy of deep learning classifiers.

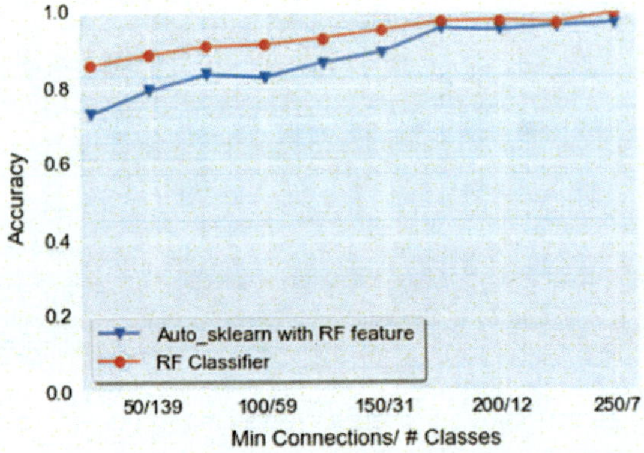

Fig. 24.2 Tenfold cross-validation accuracy on day one of data

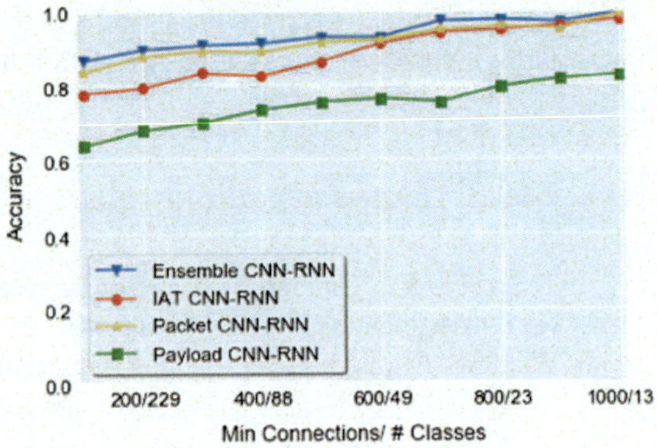

Fig. 24.3 Comparison chart of experimental results

1. Specific inputs can influence the baseline RNN's accuracy.
2. When the classification category is large, the accuracy of the baseline RNN suffers.

To address the first issue, we train the classifier using three TCP features (packet size, payload size, and arrival interval) during the handshake process. Separate classifiers are created to learn each feature of the integrated method, rather than training the three different features individually. This is done primarily because the set of these three features is relatively robust across a wide range of data streams, and training the classifier with such features will improve classification results. To address the second issue, we developed a CNN–RNN architecture and enhanced the RNN layer with more complexity and hidden units. Figure 24.1 depicts the experimental model, in which we added dropout to the arrival time CNN–RNN that tends to overfit and removed a layer of GRUs from our final model to reduce training time.

As shown in Fig. 24.3, our integrated CNN–RNN classifier outperforms the classifier trained on individual features significantly.

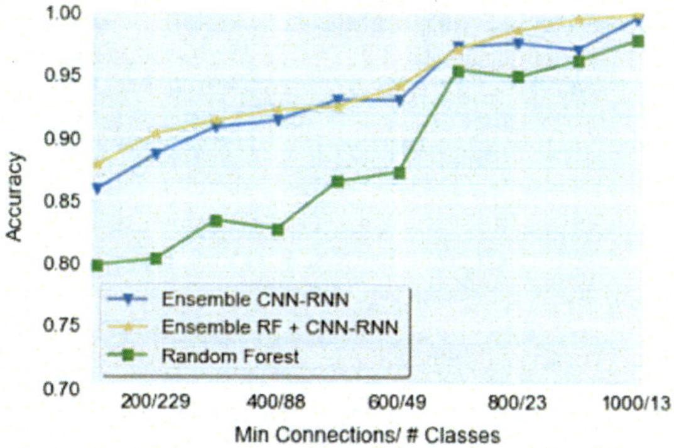

Fig. 24.4 Line graph of experimental results

After performing a tenfold cross-validation on various minimum connection thresholds, we report our final results. Because this is the most realistic and difficult scenario for a network traffic classifier, we ran our experiments with the lowest threshold values. This is also the threshold value used in for the same dataset. Using only Google Chrome data, 100 min of connections results in 532 possible SNI classes.

The tenfold cross-validation accuracy used in this experiment to improve the persuasiveness of the experimental data shows that the accuracy of the random forest classifier can hardly be improved by the deep learning architecture alone. To keep improving our classifier's accuracy, we probabilistically average the classification results of each classifier.

$$\hat{y} = \arg\max \frac{1}{4}\alpha_{\text{RF}}(x)_y + \frac{1}{4}\alpha_{\text{packet}}(x)_y + \frac{1}{4}\alpha_{\text{payload}}(x)_y + \frac{1}{4}\alpha_{\text{IAT}}(x)_y$$

where $\alpha_{\text{RF}}(x)$ denotes the random forest's output probability, $\alpha_{\text{packet}}(x)$ the output probability of the classifier trained using packet size, $\alpha_{\text{payload}}(x)$ the output probability of the classifier trained using payload, and $\alpha_{\text{IAT}}(x)$ the output result of the classifier trained using arrival time interval.

According to the improved experimental results shown in Figs. 24.4 and 24.5, the integrated CNN–RNN random forest classifier outperforms the traditional random forest classifier in classification.

24.5 Conclusion

The accuracy of encrypted traffic classification can be significantly improved by combining SNI detection with deep neural network architecture. Our study focuses on HTTPS traffic services, and our main motivation for classification is that as traffic classification accuracy improves, the problem of eavesdropping on large amounts of traffic in transit can be eliminated [11].

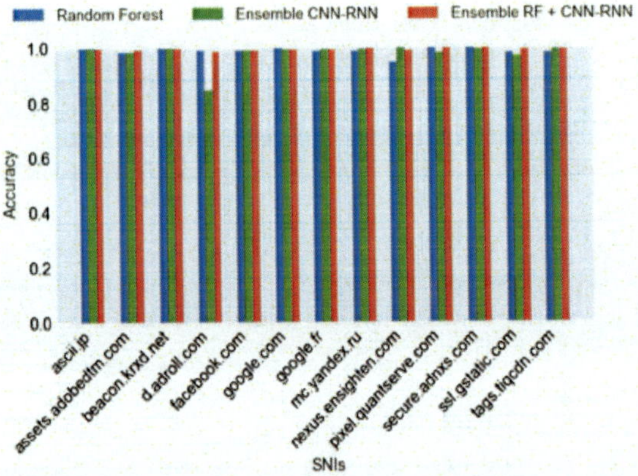

Fig. 24.5 Histogram of experimental results

For experimental studies, the classification model in this thesis primarily employs neural network architecture and incorporates random forests. We can efficiently identify SNIs in this architecture by considering only the statistics and sequences of encrypted TCP traffic, without decrypting or using any other packet header information. Our model is built with recurrent neural networks, convolutional neural networks, and random forests [12]. We achieve a high accuracy of the integrated model that, to the best of our knowledge, outperforms the latest techniques by carefully analyzing the different approaches and studying the most useful features of the network streaming data. The model will be tested on real-time HTTPS traffic in the future to see how well it predicts Internet services.

References

1. Dainotti, A., Pescape, A., Claffy, K.C.: Issues and future directions in traffic classification. IEEE Network **26**(1), 35–40 (2012)
2. Kawano, S., Okugawa, T., Yamamoto, T., Motono, T., Takagi, Y.: High-speed DPI method using multi-stage packet flow analyses. In 2012 9th Asia-Pacific Symposium on Information and Telecommunication Technologies (APSITT) (pp. 1–6). IEEE (2012).
3. Wang, W., Zhu, M., Wang, J., Zeng, X., Yang, Z.: End-to-end encrypted traffic classification with one-dimensional convolution neural networks. In 2017 IEEE international conference on intelligence and security informatics (ISI) (pp. 43–48). IEEE (2017)
4. Auld, T., Moore, A.W., Gull, S.F.: Bayesian neural networks for internet traffic classification. IEEE Trans. Neural Networks **18**(1), 223–239 (2007)
5. Moore, A. W., Zuev, D.: Internet traffic classification using bayesian analysis techniques. In Proceedings of the 2005 ACM SIGMETRICS international conference on Measurement and modeling of computer systems (pp. 50–60) (2005)
6. Draper-Gil, G., Lashkari, A. H., Mamun, M. S. I., Ghorbani, A. A.: Characterization of encrypted and vpn traffic using time-related. In Proceedings of the 2nd international conference on information systems security and privacy (ICISSP) (pp. 407–414) (2016)
7. Shen, M., Zhang, J., Zhu, L., Xu, K., Du, X., Liu, Y.: Encrypted traffic classification of decentralized applications on ethereum using feature fusion. In 2019 IEEE/ACM 27th International Symposium on Quality of Service (IWQoS) (pp. 1–10). IEEE (2019)

8. Taylor, V. F., Spolaor, R., Conti, M., Martinovic, I.: Appscanner: Automatic fingerprinting of smartphone apps from encrypted network traffic. In 2016 IEEE European Symposium on Security and Privacy (EuroS&P) (pp. 439–454). IEEE (2016)
9. Moreira, R., Rodrigues, L. F., Rosa, P. F., Aguiar, R. L., de Oliveira Silva, F.: Packet vision: a convolutional neural network approach for network traffic classification. In 2020 33rd SIBGRAPI Conference on Graphics, Patterns and Images (SIBGRAPI) (pp. 256–263). IEEE (2020)
10. Li, R., Xiao, X., Ni, S., Zheng, H., Xia, S.: Byte segment neural network for network traffic classification. In 2018 IEEE/ACM 26th International Symposium on Quality of Service (IWQoS) (pp. 1–10). IEEE (2018)
11. Wu, H., Zhang, X., Yang, J.: Deep learning-based encrypted network traffic classification and resource allocation in SDN. J. Web Eng. 2319–2334 (2021)
12. Adhao, R. B., Pachghare, V. K.: Network Traffic Classification Using Feature Selections and two-tier stacked classifier. Int. J. Next-Gener. Comput. **12**(5) (2021)

Chapter 25
Flipped Classroom Teaching Mode in College English Teaching Based on Image Recognition

Min Cheng and Jian Du

Abstract With the continuous improvement of the education recording system and artificial intelligence technology, the intelligent education system based on image recognition (IR) technology has achieved rapid development and application. The intelligent IR and analysis system for classroom behavior will analyze and record the behavior action (BA) data in the whole classroom. This work establishes an intelligent IR system using a convolutional neural network (CNN)-based image behavior action recognition algorithm. This system identifies students' behavior actions in an English flipped classroom (FC) to reflect students' motivation for the English classroom and to gauge the allure of the teacher's lecture to the students. The study's results will provide data reference in the subsequent assessment and analysis of the quality of college English teaching.

25.1 Introduction

Due to the increasing importance of education in society and the in-depth research on teaching and learning process, people's understanding of the teaching and learning process is constantly changing, and they are beginning to realize the complexity of classroom teaching behaviors. Teaching feedback is an important guarantee to optimize the teaching effect and facilitates both students' learning and teachers' teaching. Therefore, at this stage, many schools provide feedback on teaching information by identifying students' learning behavior patterns in FC teaching.

In the recent years, microlearning, FC, MOOC and other teaching modes have become a boom, and they are developing well and have achieved good results. In the research of FC teaching model in college English, some scholars study the case study conducted in English FC, in which students are assigned questions before class to read the materials and watch English videos, actively learn in class using the case study and find that students' learning ability of cooperative communication

M. Cheng (✉) · J. Du
Nanjing Xiaozhuang University, Nanjing 211171, China
e-mail: chengmincolin@163.com

S. Patnaik et al. (eds.), *3D Imaging—Multidimensional Signal Processing and Deep Learning*, Smart Innovation, Systems and Technologies 348,
https://doi.org/10.1007/978-981-99-1145-5_25

is improved, and the price to pay is that a lot of time is needed to prepare English teaching videos before class [1]. Regarding classroom behavioral IR, the current domestic recognition and analysis systems for educational scenes basically use a combination of manual control and infrared induction location tracking for human target detection and positioning. When using this method for video recording, it is necessary to equip professional personnel to constantly switch scenes and control the camera, which requires high labor cost. In addition, the infrared induction method is also prone to receive interference from external heat objects, the phenomenon of lost or false detection of target detection and tracking, which has a greater impact on the whole system [2, 3]. In this context, the intelligent image analysis system based on teaching video IR has gained more attention from developers. The recognition system based on teaching video images can save a lot of labor costs and solve classroom scene intervention during manual operations. The recognition system analyzes the teacher's behavioral actions and determines whether the teacher is writing on the board or carrying out behaviors such as lecturing and asking questions on PowerPoint lectures. The student behavior recognition module takes classroom students as subjects and identifies their behaviors in the classroom [4, 5]. It can be seen that both the research on English FC and the application of IR technology on classroom behaviors have achieved good research results.

In this paper, we first introduce the IR technology and CNN-based IR algorithm, then build an IR system for the university English FC teaching model, through which the learning behaviors of students in the FC teaching video are recognized, then test the recognition accuracy of the model on the CNN model in the IR system and finally apply the IR system to recognize the learning behavioral actions, so as to reflect the students'. Finally, we apply the IR system to identify the learning behaviors and actions, so as to reflect the students' concentration on English FC teaching.

25.2 IR Techniques and Algorithms

25.2.1 IR Technology

IR technology is to study the sensitivity of machines for IR, so that intelligent machines such as computers have certain recognition ability and free people from repetitive and heavy work [6].

IR is preceded by image processing and image feature extraction. Image processing is the process of parsing or transforming the image to be processed to obtain an accurate and clear image. The first step is to filter and remove interfering waves, de-dry, analyze grayscale values and perform image enhancement, geometric correction, etc. accordingly [7]. And then image feature extraction is performed, and how determining the appropriate feature values is one of the first key problems to be solved in designing a pattern recognition system. Extracting a certain feature value

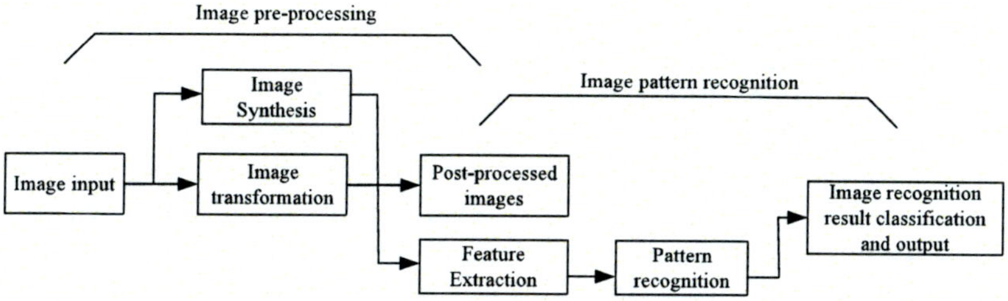

Fig. 25.1 Image processing and recognition process

from a large amount of similar object distribution data makes the feature value adaptable to the universal object distribution data; if the proposed feature value is disorganized, it cannot make the data universally adaptable. Therefore, feature extraction and analysis are particularly important steps in the whole IR [8]. Once the features are extracted, pattern matching is then performed according to the feature distribution, and finally, the classification and recognition of the image can be achieved. There is a sequence between image preprocessing and pattern recognition, but no strict boundary exists. This process can be described as shown in Fig. 25.1.

25.2.2 IR Algorithm–CNN Algorithm

In the era of immature computer technology, CNN technology was mainly used in biological research, and it has high artificial intelligence simulation ability, which can simulate biological neuron feature information and functional process more completely [9]. CNN does not need intensive training of recognition features, and it can extract feature images through feature sequences, just like the human brain can accurately discriminate eye recognition images by distinguishing individual differences. Thus, the feature information extracted by CNN is more accurate and stable, and the matchability is strong [10].

CNNs are widely used in various aspects of image processing. It can divide the image to be processed into several parts, perform feature processing such as denoising and enhancement according to the actual operation needs, retain the feature information of the original image, so as to improve the speed and accuracy of feature recognition and finally achieve the purpose of recognition detection [11].

When using CNN for IR, neurons are connected to each other, and after a certain number of iterations, the neurons in the whole image at each iteration are counted, thus producing a feature sequence and containing the features of the image to be recognized, which can well express the feature information of the image [12].

$$L(a, b) = \frac{1}{|D|} \sum_{b \in D} f(a[c], b[c]) \tag{25.1}$$

$$a[c] = \begin{cases} +1, & \text{if}\, k\|c - x\| \le R \\ -1, & \text{otherwise} \end{cases} \quad (25.2)$$

where $a[c]$ is the image label, $b[c]$ is the response map, f is the logistic loss function, k is the network step size, x is the center of the target, R is the network definition radius, L represents the number of iterations, and D represents the characteristic sequence.

25.3 Experimental Research on IR System in College English Teaching

25.3.1 University English FC Teaching Model

English FC means that English teachers create teaching videos before class, students grasp and control their own learning progress according to their own learning situation, students can watch relevant teaching videos during non-class time and learn in a relaxed atmosphere, the speed of learning video progress is under their own control and not affected by the teacher's uniform teaching progress, they can think carefully when they encounter difficulties in the learning process and return to the classroom, and teachers and students can have face-to-face communication to complete the internalization of knowledge as a way of education and realize the transfer of knowledge [13]. Students record their problems and confusions in the process of independent learning and communicate and discuss them with their peers and teachers in the classroom to solve the problems encountered in the independent learning stage. Throughout the implementation of English FC teaching, information technology and learning activities are important components of the FC, while the generation of a personalized learning environment for learners is ensured, facilitating the internalization and acquisition of knowledge for students with different foundations. The FC teaching video not only allows you to watch the teacher's teaching actions, but also records the students' actions in the background during the learning process.

25.3.2 Teaching Actions Recognition in English FC

This paper establishes an intelligent IR system that can be used to identify teacher and student behavioral actions or postures in FC teaching videos. For student behavioral action recognition, students can be recognized to stand up, sit down, hold their cheeks, lower their heads and so on. For teacher behavior, it can recognize teacher walking, boarding and other actions. This system uses an intelligent tracking camera to track the teacher and always keep a close-up view of the teacher to ensure detailed information of the teacher's lecture in the classroom [14].

25.3.3 Hardware Platform of FC Behavior Intelligent IR System

Hi3531A chip is used as the main chip of the classroom behavior intelligent IR and analysis system, which has rich interface resources. It can be configured with some peripheral chips to complete the work of audio and video acquisition, processing, analysis and storage. Based on the Hi3531A main chip, the hardware part of the intelligent IR system is built.

25.3.4 IR System Module of FC Teaching Video

As shown in Fig. 25.2 is the module structure of the IR system, and the module functions are described as follows.

Image information acquisition module. The camera captures the video images, and the image information acquisition module is mainly to read the images captured by the camera into the system. English teachers usually use webcam when creating teaching videos. To get the content of the webcam, they need to be in the same LAN with the webcam, and then use OpenCV to get the data information of the webcam. The camera acquires a uniform image size, which ensures that the deformation of the features contained in the image is also uniform when scaled to a fixed size, which will facilitate the efficiency of recognition.

Image information processing module. The main function of this module is to convert the teaching image information into a coded form that can be used as input data for the behavioral action recognition module by first processing it once. This module should also eliminate useless information and minimize the complexity of the data while ensuring that key information is not lost, in order to reduce the amount of

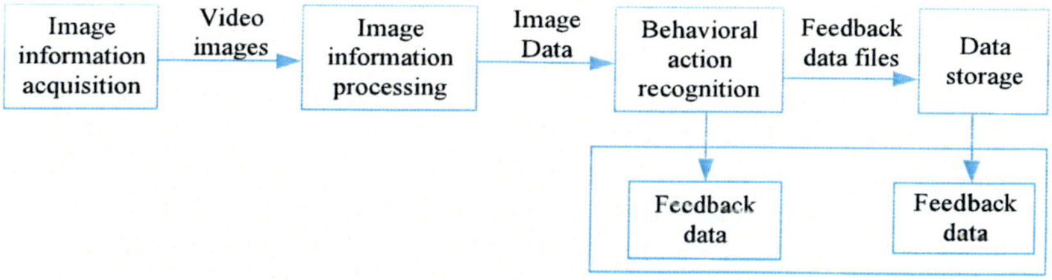

Fig. 25.2 System module relationship diagram

computation in the system and allow the system to have higher operational efficiency. For detecting human body and recognizing human behavioral actions, the information contained in the color of the image does not have practical significance. The gray image and the color image contain basically the same effective feature information, and transforming the image from color to gray does not lose any human feature information. Therefore, the first job of the image information processing module is to grayscale the image to reduce the system's computational overhead.

BA recognition module. BA recognition module is the core module of the system, which extracts the teaching feedback data from the teaching video images, and all the functions of the system are based on these teaching feedback data. The BA recognition module needs a neural network (CNN) to complete the detection and recognition of images. The core of this module is a trained neural network model with some computational functions of indicator data.

Data storage module. The function of this module is to persist the data to be used for the calculation and save it in the hard disk. The real-time data output in the BA recognition module is written to the file of the computer where it is located, and the data display function of the server needs to access these files, so it is also necessary to save these files uniformly. Unlike relational databases, the system generates unstructured data, so HDFS is chosen to store these files.

25.4 Experimental Analysis

25.4.1 Testing of CNN Model

An English teaching video is used to test the effect of CNN model recognition. Because the amplitude of action change is very small between consecutive frames of the video, one frame is extracted from every ten frames of the image, and the video is split into 200 images, which have a larger amplitude of action change than the normal consecutive frames, and the variability of the image content is also greater. These 200 images are all images that contain targets, and all targets cover the full range of behavioral poses. In the process of inspection, these 200 images were sequentially input into the CNN model for recognition, the resultant images were recognized by human judgment of accuracy and statistics, and the results are shown in Table 25.1.

Where the real targets are all the targets in the image. Detected targets are all the borders generated by the model prediction, and missed targets are the real targets that the model does not detect. The detected wrong targets are the targets detected by the model. After inspection, the CNN model achieved 85.0% accuracy on the tested data, but there was also 18.83% omission. The condition causes that the training samples are not abundant enough. With abundant training samples and sufficient training time, better accuracy and higher generalization ability can be achieved. The model can be integrated into the system after the training is completed.

Table 25.1 Model detection results

Item	Number of targets
Real targets	754
Detected targets	513
Missing targets	142
Correctly detected targets	436
Incorrectly detected targets	80
Detection omission rate	18.83%
Detection accuracy rate	85.0%

25.4.2 Analysis of Behavioral Action Recognition in College English FC

A simple test experiment is conducted on the IR system using the English teaching video with the above-tested model. The camera path of the image information acquisition module is set to point to the video file, and then the client program is run. Using the IR system to recognize the behavioral actions of students in the teaching video, including watching the video, reading and writing, resting their cheeks, bowing their heads, lying down and other actions, the test results are shown in Fig. 25.3.

The mean, plural, maximum and minimum values of each behavioral action identified in Fig. 25.3 were used to calculate students' concentration on the English FC teaching mode. The results are shown in Table 25.2.

Since fewer than 5 people were in the test video, many of the categorized action behaviors were not included most of the time. In the test results, the mean and plural dimensions for English FC concentration scores were above 80 for a positive classroom. In the instructional videos, students change from one type of action pose

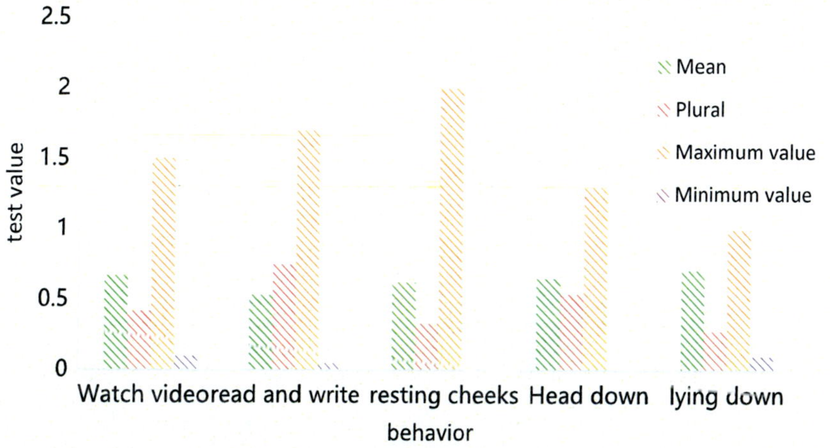

Fig. 25.3 Results of the behavioral action recognition test of teaching video

Table 25.2 Students' concentration in class

	Mean	Plural	Maximum value	Minimum value
Concentration	87.00	84.00	96.00	0.00

to another, such as watching video pose to hand on cheek pose, and the image frames captured in this action are more difficult to define their classification, causing fluctuations in the detection results are inevitable. To further improve the accuracy of system recognition, more and more complex samples need to be produced for training in order to make the boundaries of action recognition classification clearer. From the experimental results, it seems that the recognition accuracy of the current system, with the average value of concentration as the reference index, can achieve the purpose of the system and provide a reference for the evaluation work of English FC teaching by teaching managers.

25.5 Conclusion

In the English FC teaching environment, students' learning status can be reflected in posture, demeanor, facial expressions, etc., and behavioral action posture is the most intuitive and easily accessible information. In this paper, IR technology is applied to the FC teaching environment to recognize students' behavioral actions in order to collect teaching feedback information. The CNN model is introduced in the key IR system, and by training the CNN model, it is found that the model has a correct high rate for IR detection and is suitable for IR of instructional videos. In addition, the IR system is used to analyze and identify students' learning behaviors in English flipped classroom and calculate the students' concentration level on English FC.

References

1. Tsai, T.H., Chi, P.T.: A single-stage face detection and face recognition deep neural network based on feature pyramid and triplet loss. IET Image Proc. **16**, 2148–2156 (2022)
2. Shaik, J. B., VS, A., Singhal, S., Goel, N.: Reliability-aware design of temporal neuromorphic encoder for image recognition. Int. J. Circuit Theory Appl. 50, 1130–1142 (2022)
3. Naosekpam, V., Sahu, N.: Text detection, recognition, and script identification in natural scene images: a review. Int. J. Multimed. Inf. Retr. 1–24 (2022)
4. Favorskaya, M.N., Pakhirka, A.I.: Image-based anomaly detection using CNN cues generalisation in face recognition system. Int. J. Reason.-Based Intell. Syst. **14**, 19–26 (2022)
5. Hashim, B., Amutha, R.: Deep transfer learning based human activity recognition by transforming IMU data to image domain using novel activity image creation method. J. Intell. Fuzzy Syst. 1–8 (2022)
6. Behmanesh, F., Bakouei, F., Nikpour, M., Parvaneh, M.: Comparing the effects of traditional teaching and flipped classroom methods on midwifery students' practical learning: the embedded mixed method. Technol., Knowl. Learn. (2020)

7. Gren, L.: A Flipped Classroom Approach to Teaching Empirical Software Engineering. IEEE Trans. Educ. 1–9 (2020)
8. Kay, R.H., Macdonald, T., Digiuseppe, M.: A Comparison of lecture-based, active, and flipped classroom teaching approaches in higher education. J. Comput. High. Educ. **31**, 449–471 (2019)
9. Abdullah, M.Y., Hussin, S., Ismail, K.: Does flipped classroom model affect EFL learners' anxiety in English speaking performance? Supyan Hussin, Kemboja Ismail. Int. J. Emerg. Technol. Learn. (iJET) **16**, 94–107 (2021)
10. Abe, Y., Elwood, J. A., Yin, Y. K., Hood, M.: The relationship between self-regulation skills and English proficiency among Asian EFL learners in the flipped online classroom. Int. J. Knowl. Learn. 14 (2021)
11. Eppard, J., Rodjan-Helder, M.G.D., Baroudi, S., Reddy, P.: Integrating flipped learning into an english pre-sessional class at a public university in the UAE: reports from an SLL university classroom. Int. J. Virtual Pers. Learn. Environ. (IJVPLE) **11**, 65–86 (2021)
12. Akayoğlu, S.: Teaching CALL to pre-service teachers of English in a flipped classroom. Technol. Knowl. Learn. **26**, 155–171 (2021)
13. Tokmak, H.S., Yakin, I., Dogusoy, B.: Prospective English teachers' digital storytelling experiences through a flipped classroom approach. Int. J. Distance Educ. Technol. (IJDET) **17**, 78–99 (2019)
14. Abdullah, M. Y., Hussin, S., Ismail, K.: Implementation of flipped classroom model and its effectiveness on english speaking performance. Int. J. Emerg. Technol. Learn. 14 (2019)

Chapter 26
Computer-Aided Design and Furniture Design Practice Research

Jing Zeng

Abstract In the trend of economic globalization, in the face of the rapid development of material technology, information resources, communication platform and other contents, the external form of manufacturing industry has undergone fundamental changes, from fast delivery to fast response to market demand, and has become the core of market competition. Especially for furniture design and manufacturing industry, in order to better cope with the opportunities and challenges brought by the development of the new era, it is necessary to develop application product computer-aided design system on the basis of integrating previous development experience, so as to improve its comprehensive development level. Therefore, on the basis of understanding the development status of furniture design and computer-aided technology in the recent years, according to the CAD system and related functions in the field of furniture design, this paper deeply discusses the design idea and overall structure of the product life cycle information model as the core, so as to provide technical support for the development of modern furniture manufacturing industry.

26.1 Introduction

Furniture design in the traditional sense is mainly based on tools such as drawing board, rubber, ruler and pen, integrating design and trial production. When designing to make furniture product, need to pass a lot of steps, for instance model conception, draw sketch, selected plan, make model, modify model, product finalize the design, custom-made production. In this process, affected by internal and external factors, increase product design cycle.

For example, in product lighting design, it is necessary to consider the intensity of the light source that has a great influence on the use of space characteristics and consider the lighting degree of the light source and visual sense. The ability of visual senses to identify space will be reduced to the minimum, which indirectly affects

J. Zeng (✉)
Arts College of Sichuan University, Chengdu, China
e-mail: 13976@sicau.edu.cn

S. Patnaik et al. (eds.), *3D Imaging—Multidimensional Signal Processing and Deep Learning*, Smart Innovation, Systems and Technologies 348,
https://doi.org/10.1007/978-981-99-1145-5_26

their psychological perception of space. The improvement of illumination degree will affect spatial recognition ability and visual psychological perception to a certain extent. The adaptability of visual physiology and the economic benefit of light source are considered. Therefore, without considering physical visibility and psychological perception, designers often adopt a variety of lighting methods and techniques to change the contrast between light and shade of space, stimulate the visual sense of users and then produce different visual perception of space, emphasizing the nature and purpose of space [1].

And computer-aided design technology is the use of computer's fast computing ability, processing complex and changeable painting and calculation work, so that designers have enough time and energy to create ideas and product design. Combined with relevant product factors, the effective combination of artificial intelligence and computer can deeply inspire designers' creative thinking and inspiration, liberate them from the traditional monotonous and complex working mode and give full play to their creative potential. Autodesk has successively introduced enhanced AutoCAD and 3DVIZ software versions, providing excellent secondary development platform and animation design software for designers. Among them, the accurate graphic design of AutoCAD and the powerful animation function of 3DVIZ software provide a new development environment for modern furniture designers. For example, the secondary development based on AutoCAD refers to the combination of different development environments, the use of different development languages, according to the library files and header files provided by AutoCAD software interface program, and thus process the original parameter data, so that AutoCAD software can automatically or semi-automatically perform repetitive calculation and drawing tasks. In order to improve the efficiency of furniture design. From the point of view of practical application, when CAD technology is designing furniture products, the general picture library of furniture parts will be constructed first, and then according to the design requirements of products, the corresponding parts will be obtained from the picture library, the program that conforms to the design requirements will be selected in respective programs, and the cost accounting will be completed finally. In the new era of technological innovation and development, computer-aided furniture design can let the staff get rid of the limitation of the drawing board comprehensively, while reducing the working pressure, improving the speed of the drawing and ensuring the quality of the design drawings.

At the same time, AutoCAD software can also provide designers with a variety of graphics, textures, colors, so that designers on the product function and visual effects more vivid. Nowadays, FCAD software has been widely used in most furniture manufacturing enterprises, which indicates that the development of science and technology will inevitably accelerate the changes in the production field [2, 3].

Based on the systematic understanding of the development trend of furniture design in China in the recent years, it can be seen that the actual consumption concept tends to be more personalized, and consumers begin to pursue the appreciation and taste of personal rooms. Among them, the people-oriented green furniture design concept has become the main direction of furniture design development in the new era. This not only requires designers to combine the personalized characteristics of human beings and the characteristics of housing construction to carry out deep thinking, but also is based on the development requirements of ecological industry, protection, development and rational use of natural materials. From the perspective of the development trend of economic globalization, many furniture enterprises are discussing how to adapt to the market demand of small batch and many varieties, so the information construction of furniture industry is urgent. Although there are many problems at present, China's furniture design, such as lack of innovation, design and material single, furniture products are the same, but with the continuous development of modern information technology, computer-aided technology, widely used in the pattern and to optimize the design of the furniture, and to provide consumers with individuation and diversity of furniture products. Therefore, this paper mainly studies the AutoCAD software as the core of computer-aided furniture design [4].

26.2 Method

26.2.1 AutoCAD Software

As a powerful visual application development tool, Visual C++ can help Windows programmers to use the most dynamic professional development environment, so it is also considered as Windows 32-bit platform basic application. At the same time, the powerful AppWizard program generator can reduce the amount of programmer's work and reduce the probability of errors in the program. Combined with the AutoCAD software flowchart analysis shown in Fig. 26.1, AutoCAD will regard Visual C++ as its own development environment [5].

In the furniture design part, designers should not only consider the furniture product itself, but also are based on function and esthetics, to the entire interior space effective extension and comprehensive reflection. In the development of intelligent.

In the intelligent software development, the designer will carry out the steps of design, production and sales of furniture information and apply the steps of market information, user feedback and production quality information in the design work, so that flexible and efficient design software can be used to reduce the excessive consumption of unnecessary resources. From the perspective of practical application, furniture design in 1AutoCAD software contains three contents. First, it refers to furniture modeling design; second, it refers to furniture color design; and finally, it refers to furniture structure design. Among them, in furniture modeling design, designers want to consider the content such as the appearance, dimension and integral structure of furniture according to different type key [6].

Taking the calculation of spare parts size as an example, the calculation formula of single edition can be clearly described by introducing parametric and formulaic design concepts in furniture design, and according to the matching relationship between cabinet parts and hardware. This feature is very suitable for the application

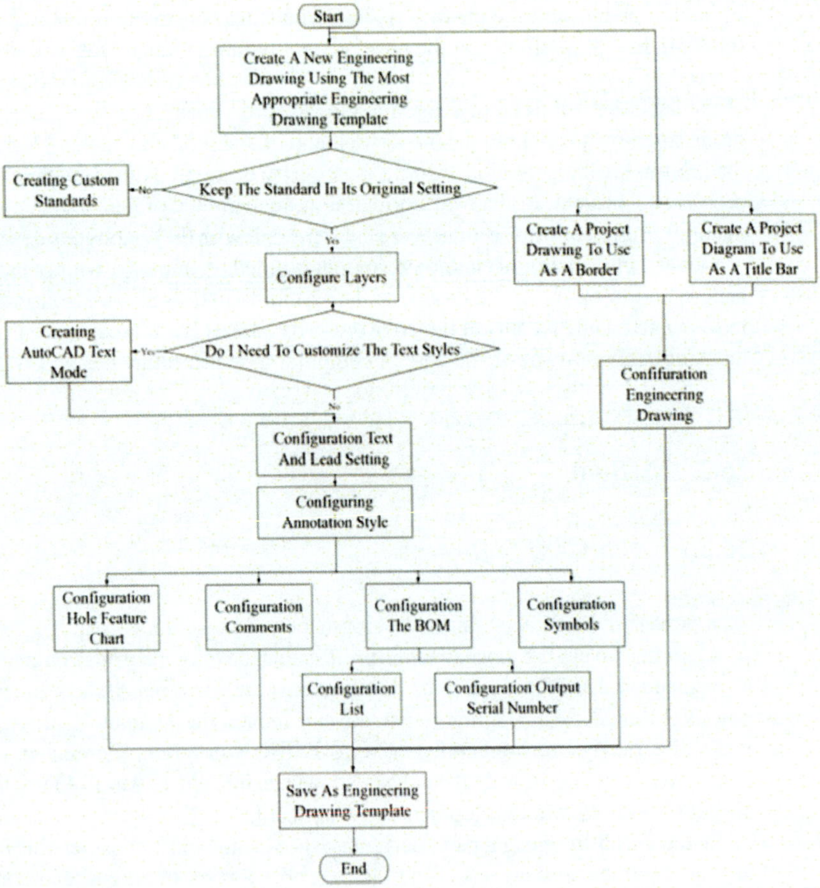

Fig. 26.1 AutoCAD software flowchart

of computer-aided design, mainly combined with human thinking, group description, computer standard operation and so on to achieve the combination of units, thus forming a variety of styles of furniture world. Table 26.1 below is the formula for calculating the specific parts size. See Table 26.1 for details.

26.2.2 Furniture Design Process

Based on the analysis of the design flowchart of the furniture shown in Fig. 26.2, it can be seen that there are four stages.

Firstly, the problem formation stage. This stage is based on the market research, after mastering the existing design furniture products, in-depth research on their

Table 26.1 Dimension calculation formula of parts

The name of the panel		Dimension calculation formula	Note
Side plate	The X direction	18	The thickness of the plate is 18 mm
	Y direction	L	
	The Z direction	H	
Roof	The X direction	W – 36 – 1	W is the width of the cabinet body, and 1 is the margin
	Y direction	L	
	The Z direction	18	
The back	The X direction	W + 4 + 4	4 Insert the groove depth for the backplane
	Y direction	5	Plywood thickness
	The Z direction	H – 100 – 18 + 4 + 4	4 Indicates the depth of the slot inserted into the backplane
Floor	The X direction	W – 36 – 1	W is the width of the cabinet body, and 1 is the margin
	Y direction	L	
	The Z direction	18	
Vertical partition	The X direction	18	
	Y direction	L	
	The Z direction	D – 1	D is the net height of the cabinet, and 1 is the margin
Diaphragm plate	The X direction	S – 1	S is the net width of space, and 1 is the margin
	Y direction	L	
	The Z direction	18	
Plastic plate	The X direction	W – 1	W is the width of the cabinet body, and 1 is the margin
	Y direction	18	
	The Z direction	100	
The door	The X direction	S + 7 + 7	S is the net width of space, and 1 is the external lap
	Y direction	18	

(continued)

Table 26.1 (continued)

The name of the panel		Dimension calculation formula	Note
	The Z direction	D + 7 + 7	D is the net height of the cabinet, and 7 is the amount of outside
Hang garment lever	The X direction	S – 1	S is the net width of space, and 7 is the external lap
	Y direction	R	R is the radius of the hanging rod
	The Z direction	R	R is the radius of the hanging rod

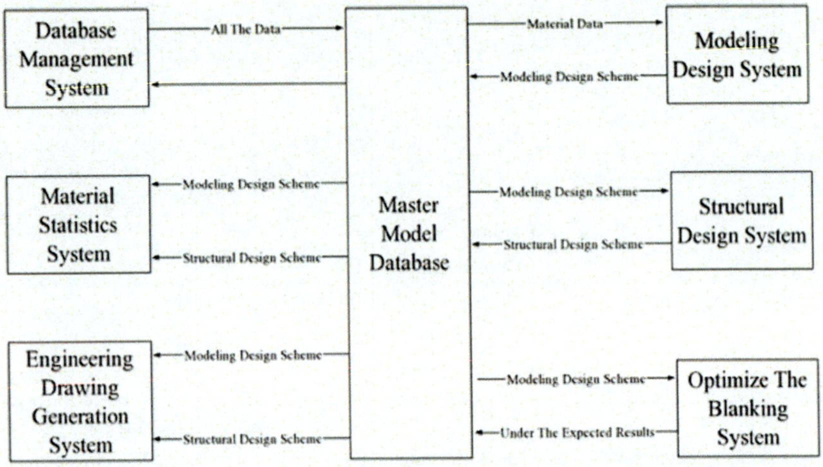

Fig. 26.2 Flowchart of home design

advantages and disadvantages and comparative analysis of similar products or imitation products, so as to get the core interests and essential functions of the new products and clear the market objectives and price positioning of furniture products. The results of this phase are reflected in product development information or images.

Secondly, the problem-solving stage. This stage belongs to the idea of the product design, according to the design objectives proposed in advance, deep discussion and solve the relevant problems. For example, the detailed processing of customer needs, in-depth discussion of product applicability, reliability, maintainability, durability, safety and other basic requirements, and according to relevant standards and rules, put forward effective ways to use new technology [7].

Thirdly, design result transformation stage. This stage includes drawing, model, scheme, trial production, market and other processes, belonging to formal product design and product planning. Designers should present the functional form, functional size, modeling color, material structure and other contents of furniture products in detail and then determine the plan, transform the relevant drawings into specific products and finally become required goods through the market. In this stage, it also includes additional content such as sales plan and after-sale service for promoting products.

Finally, information feedback stage. At this stage, it is necessary to collect customers' opinions and find out the problems existing in all aspects of the new product and the problems of sales and after-sales service. Through extensive investigation and sorted into written documents, it can be used as the main basis for subsequent furniture product design [8].

26.3 Result Analysis

26.3.1 System Applications

In the rapid development of modern science and technology, furniture production and design emerged new characteristics, such as technology tends to cooperate and automation, the selection of product parts with standardized and standardized characteristics, style design more attention to personalized and natural. Although furniture enterprise and design personnel have realized the application value of CAD technology, but in practical work still give priority to with manual design, this and other industry compare, will produce bigger gap. Nowadays, with the continuous improvement of furniture CAD commercialization software in China, facing higher and higher requirements of furniture design and computer-aided design, the development of panel furniture CAD system, also known as FCAD. The overall architecture of the system is shown in Fig. 26.3 [9].

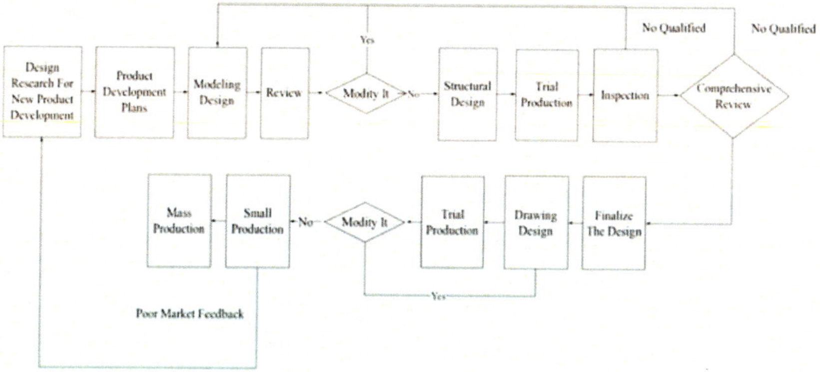

Fig. 26.3 System architecture diagram

From the perspective of practical application, Visual C++ programming is the main part of the application of this system, which effectively integrates AutoCAD and other technical software [10]. As long as designers enter this system, they can complete all operations of panel furniture design, such as drawing, generating reports and data calculation. From the perspective of application performance, this system can automatically call collaborative software to complete various design tasks, avoiding designers to switch back and forth between different software systems [11]. At the same time, all subsystems do not directly communicate with each other, but exchange data through the master template database, which can not only reduce the coupling degree of system modules, but also improve the robustness of the system and simplify the steps of furniture design [12].

26.3.2 Development Trend

With the rapid development of information technology, computer-aided design (CAD), as shown in Fig. 26.4, has been widely used in many fields. It has not only changed the traditional design, production and organization mode, but also further accelerated the development of furniture. Combined with the above CAD software technology analysis, it can be seen that the application function and basic performance of furniture product design in computer-aided design can further improve the efficiency and quality of furniture design and production. For example, in 3D modeling, complex product designs can be transformed into simple part modeling. Among them, the three-dimensional model, the establishment method is divided into two kinds, one is based on vertebral body, cylinder, cube, using intersection, union, difference and Boolean operation to generate complex model; the other is to build a model based on sketches and select modified features in combination with basic features. Both can generate real and unique three-dimensional entities, which are the focus of product design in the field of furniture [13].

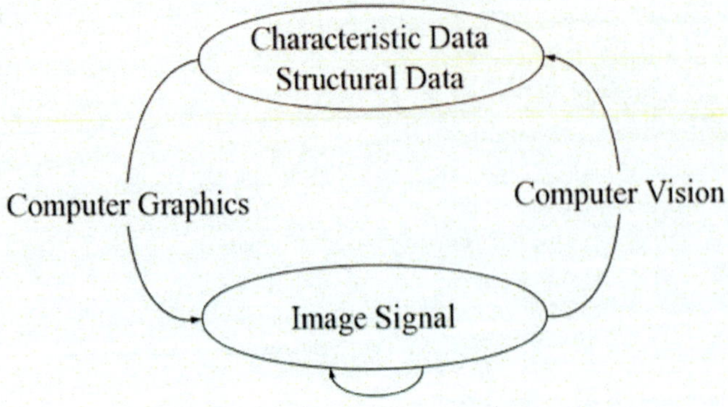

Fig. 26.4 Computer-aided design structure diagram

26.4 Conclusion

To sum up, CAD software system, as a powerful and widely applied new technology, is widely used in the field of furniture. It can not only fully show the application value of computer-aided design, but also optimize the design level of furniture products in practical exploration. Therefore, Chinese scholars should be in Chinese style furniture design and computer-aided design of the research at the same time, actively learn from foreign advanced research results, combine with the new era under the furniture in the field of product design requirements, present diversity and personalized products with good quality, can meet the demand of consumers to buy this and continuously optimize the technology development of China's furniture industry level [14, 15].

Acknowledgements This work was supported by "Sichuan key research base of philosophy and social sciences-center of modern design and culture", Project Number "MD19E006".

References

1. Huang, C.H.: Application of computer-aided design in modern furniture–comment on micro-course graphic tutorial of furniture computer-aided design. Wood Ind **34**(5), 1 (2020)
2. Xue, X.T.: Teaching reform and practice of autoCAD software in furniture design specialty based on the background of "integration of industry and education". Sci. Technol. Wind (2), 3 (2022)
3. Kang, Y.E.: Design of curriculum framework of computer-aided design for urban planning management. Fujian Light Textile (6), 4 (2022)
4. Yu, Y., Yu, H. J., Wang, X. N., et al.: Research on multi-situation teaching mode of computer-aided design and manufacturing. Teach. Educ. High. Educ. Forum (5), 3 (2020)
5. Wang, Y.Q., Zheng, G.H., Wang, Y.H., et al.: Current situation and prospect of computer aided design (CAD) and virtual reality technology in concentrator. Gold Sci. Technol. **29**(5), 11 (2021)
6. He, M.F., Yu, J. H.: Discussion on curriculum system reconstruction of computer-aided design for industrial design major (2017–11), 53–53 (2021)
7. Chen, F.H.: Computer-aided design and model making (I) application curriculum reform exploration. J. Chin. Multimed. Netw. Teach. **02**, 151–152 (2020)
8. Yang, X.Y.: Research on the combination of mechanical drawing course and computer-aided design course teaching reform. Lit. Youth **11**, 0234–0234 (2021)
9. Zhou, J.: Application of numerical control technology and computer-aided design in numerical control machine tool. Equip. Manag. Maint. (18), 3(2020)
10. Guo, T.S.: Application of numerical control technology and computer-aided design in numerical control machine tool. Digital Technol. Appl. **38**(2), 2 (2020)
11. Gligorović, B., Desnica, E., Palinkaš, I.: The importance of ergonomics in schools–secondary technical school students' opinion on the comfort of furniture in the classroom for computer aided design. In: IOP Conference Series: Materials Science and Engineering, vol. 393, No. 1, p. 012111. IOP Publishing (2018)
12. Molnar, J.A.: Integrated computer aided design practices as demonstrated on a fin-line device. IETE J. Res. **41**(1), 39–44 (1995)
13. Šedivý, J., Hubalovsky, S.: Principles of computer aided design in practice research. In Applied Mechanics and Materials, vol. 336, pp. 1370–1373. Trans Tech Publications Ltd. (2013)

14. Holbach, G.: Special issue on computer-aided ship design: some recent results and steps ahead in theory, methodology and practice dedicated to professor Horst Nowacki on the occasion of his 75th birthday. Comput. Aided Des. J. Comput. Aided Des. (JCAD) **42**(11), 953–955 (2010)
15. Malhotra, M.K., Heine, M.L., Grover, V.: An evaluation of the relationship between management practices and computer aided design technology. J. Oper. Manag. **19**(3), 307–333 (2001)

Author Index

S. Patnaik et al. (eds.), *3D Imaging—Multidimensional Signal Processing and Deep Learning*, Smart Innovation, Systems and Technologies 348,
https://doi.org/10.1007/978-981-99-1145-5